AF875945

Kai Melde

The Acoustic Hologram and Particle Manipulation with Structured Acoustic Fields

The Acoustic Hologram and Particle Manipulation with Structured Acoustic Fields

by
Kai Melde

KIT Scientific Publishing

Dissertation, Karlsruher Institut für Technologie
KIT-Fakultät für Maschinenbau

Tag der mündlichen Prüfung: 29. Mai 2019
Hauptreferent: Prof. Dr. Jan Gerrit Korvink
Korreferent: Prof. Dr. Peer Fischer
Prüfungsvorsitz: Prof. Dr.-Ing. Wolfgang Seemann

Die Arbeit entstand am Max-Planck-Institut
für Intelligente Systeme, Stuttgart
Forschungsgruppe Mikro-, Nano- und Molekulare Systeme

Impressum

KIT Scientific Publishing

Karlsruher Institut für Technologie (KIT)
KIT Scientific Publishing
Straße am Forum 2
D-76131 Karlsruhe

KIT Scientific Publishing is a registered trademark
of Karlsruhe Institute of Technology.
Reprint using the book cover is not allowed.

www.ksp.kit.edu

This document – excluding the cover, pictures and graphs – is licensed under a Creative Commons Attribution-Share Alike 4.0 International License (CC BY-SA 4.0): https://creativecommons.org/licenses/by-sa/4.0/deed.en

The cover page is licensed under a Creative Commons Attribution-No Derivatives 4.0 International License (CC BY-ND 4.0): https://creativecommons.org/licenses/by-nd/4.0/deed.en

Print on Demand 2019 – Gedruckt auf FSC-zertifiziertem Papier

ISBN 978-3-7315-0946-2
DOI 10.5445/KSP/1000096129

The Acoustic Hologram and Particle Manipulation with Structured Acoustic Fields

Zur Erlangung des akademischen Grades eines
DOKTORS DER INGENIEURWISSENSCHAFTEN (Dr.-Ing.)

von der KIT-Fakultät für Maschinenbau des
Karlsruher Instituts für Technologie (KIT)
angenommene

DISSERTATION

von

Dipl.-Ing. Kai Melde

Tag der mündlichen Prüfung: 29.05.2019

Hauptreferent: Prof. Dr. Jan Gerrit Korvink
Korreferent: Prof. Dr. Peer Fischer
Prüfungsvorsitz: Prof. Dr.-Ing. Wolfgang Seemann

Abstract

This thesis presents holograms as a novel approach to create arbitrary ultrasound fields. It is shown how any wavefront can simply be encoded in the thickness profile of a phase plate. Contemporary 3D-printers enable fabrication of structured surfaces with feature sizes corresponding to wavelengths of ultrasound up to 7.5 MHz in water—covering the majority of medical and industrial applications. The whole workflow for designing and creating acoustic holograms has been developed and is presented in this thesis. To reconstruct the encoded fields a single transducer element is sufficient. Arbitrary fields are demonstrated in transmission and reflection configurations in water and air and validated by extensive hydrophone scans. To complement these time-consuming measurements a new approach, based on thermography, is presented, which enables volumetric sound field scans in just a few seconds. Several original experiments demonstrate the advantages of using acoustic holograms for particle manipulation. Most notably, directed parallel assembly of microparticles in the shape of a projected acoustic image has been shown and extended to a fabrication method by fusing the particles in a polymerization reaction. Further, seemingly dynamic propulsion from a static hologram is demonstrated by controlling the phase gradient along a projected track. The necessary complexity to create ultrasound fields with set amplitude and phase distributions is easily managed using acoustic holograms. The acoustic hologram is a simple and cost-effective tool for shaping ultrasound fields with high-fidelity. It is expected to have an impact in many applications where ultrasound is employed.

Zusammenfassung

Ein wesentlicher Beitrag dieser Arbeit ist das *akustische Hologramm* als neues Werkzeug zur Formung von Schallfeldern.[1,2†] Dieses Hologramm kann als Beugungselement realisiert werden, wobei die Wellenform im Dickenprofil einer flachen Platte kodiert ist. Eine durchlaufende Schallwelle wird dabei abhängig von der lokalen Dicke der Platte verzögert. Damit ist dieses Beugungselement vergleichbar mit der Kinoform in der Optik.[3] Im Zuge der vorliegenden Arbeit wurden alle nötigen Arbeitsschritte entwickelt: von der Berechnung der Hologramme, über die physische Implementierung bis zur experimentellen Feldrekonstruktion. Dabei konnten viele Erkenntnisse und Methoden aus der optischen Holographie übernommen werden. Mit diesem neuen akustischen Bauelement wurden schließlich verschiedene Methoden der Partikelmanipulation umgesetzt und erweitert.

Holographie basiert auf der Erkenntnis, dass mit der Wellengleichung die gesamte Entwicklung eines Wellenfeldes in einem Raum durch die Kenntnis der Feldvariablen (Schalldruck und -schnelle in der Akustik) auf einer umschließenden Fläche bestimmt ist.[4,5] Unter bestimmten Einschränkungen kann diese Fläche auf die kleinere Fläche einer Apertur begrenzt werden. Kontrolle über die Amplitude und vor allem der komplexen Phase der durchschreitenden Welle in diesem Bereich ermöglicht damit Kontrolle über das gesamte Feld im anschließenden Volumen.[6] In den Anfängen der Holographie wurden die Feldvariablen für ein abzubildendes Objekt mit speziellen holographischen Methoden aufgezeichnet. Die gespeicherte Form wird schließlich als Hologramm

[†] Verweise auf Quellen werden in dieser Arbeit durch hochgestellte Ziffern gekennzeichnet (Quellenverzeichnis, S. 101ff).

bezeichnet und erlaubt die Rekonstruktion eines virtuellen oder reellen Bildes inklusive Tiefeninformation.[7] Als digitale Holographie bezeichnet man Verfahren, bei denen die Hologramme für beliebige Bilder direkt berechnet werden. Da das inverse Problem der Hologrammberechnung bei gegebener Feldverteilung jedoch schlecht gestellt ist, und die technischen Möglichkeiten der Modulation und Erzeugung beliebiger Wellen beschränkt sind, werden Hologramme in der Regel durch Optimierungsverfahren angenähert.[8] Im Zuge dieser Arbeit wurde eines dieser Verfahren, das iterative angular spectrum approach (IASA), welches für Berechnungen im Nahfeld ausgelegt ist,[9] auf die Akustik übertragen. Damit konnten akustische Hologramme für beliebige zweidimensionale Schalldruckbilder berechnet werden. Auch die Kodierung drei-dimensionaler Schalldruckverteilungen konnte umgesetzt werden, jedoch vereinfacht durch die gleichzeitige Erzeugung zwei-dimensionaler Schallbilder in separaten Bildebenen. Weiterhin wurde mit dieser Arbeit gezeigt, wie zusätzlich zum Schalldruck auch die komplexe Phase in einem Schallbild definiert und in einem Hologramm gespeichert werden kann. Die aufwändigen Oberflächenprofile der akustischen Hologramme lassen sich besonders einfach durch additive Fertigung (3D-Druck) herstellen. Die für diese Arbeit gefertigten Hologramme entsprechen der Modulation einer Welle mit bis zu 15.000 individuellen Pixeln.[1] Damit liegen sie um eine bis zwei Größenordnungen über konventionellen Phasenwandlern. Handelsübliche Drucker ermöglichen zur Zeit die Herstellung beugungslimitierter Hologramme für Ultraschall bis 7.5 MHz in Wasser. Dieser Bereich deckt bereits den Großteil der technischen und medizinischen Anwendungen des Ultraschalls ab.[10,11] Höhere Frequenzbereiche können sicher durch aufwändigere Methoden der Mikrofabrikation erschlossen werden.

Messungen mit einem kalibrierten Hydrophon bestätigten dann, wie gut die experimentell rekonstruierten Felder mit den simulierten Feldern übereinstimmen. Hydrophone sind präzise Instrumente, zeichnen aber nur den Schalldruck in einem Punkt auf und werden daher seriell durch das Feld gerastert. Die Vermessung ganzer Bildebenen ist folglich sehr zeitintensiv und benötigt meist mehrere Stunden, in Extremfällen auch Tage. Aus diesem Grund wurde ein ergänzendes *thermographisches Messverfahren* entwickelt, welches die Schallverteilung in einem Volumen in wenigen Sekunden erfassen kann.[12] Dabei wird eine schalldämpfende Membran durch das Feld bewegt, während eine Wärmekamera kontinuierlich ihre Oberflächentemperatur aufzeichnet. Mit diesen Wärmebilddaten, und vorheriger Kalibrierung durch ein bekanntes Feld, lässt sich die Druckverteilung im gesamten Raum ermitteln.

Dass Schallwellen auch Kräfte übertragen können, ist lange bekannt,[13] und wird seit einigen Jahren zur Partikelmanipulation, zum Beispiel in der Mikrofluidik, eingesetzt. Jedoch basieren bisherige Ansätze auf Resonatorgeometrien mit entsprechend symmetrischen Schallfeldern und ebenso symmetrischen Partikelanordnungen.[14] Akustische Hologramme ermöglichen weitaus mehr Kontrolle über die Verteilung des Schalldrucks. Damit konnte schließlich die *erste Partikelassemblierung* in Form eines beliebigen projizierten Schallbildes demonstriert werden.[1] Diese Partikelgefüge werden durch die akustische Strahlungskraft in einem Wasserbad zusammengehalten und zerfallen, wenn der Schallsender verstummt. Die meisten Feststoffe weisen, in Wasser suspendiert, durch ihre Festigkeit und Dichte einen positiven akustischen Kontrast auf.[15] Dadurch werden sie aus den Druckmaxima verdrängt. Der Einfachheit halber wurden in den vorliegenden Experimenten daher Mikropartikel aus Silikon verwendet. Deren erhöhte Kompressibilität im Vergleich zu Wasser entspricht einem negativen akustischen Kontrast und die Partikel ordnen sich folglich mit dem positiven Druckbild an. Um die assemblierten Gefüge zu verfestigen wurden die Silikonpartikel funktionalisiert und dabei Photoinitiator auf ihrer Oberfläche aufgebracht. Nach erfolgter akustischer Anordnung der Partikel in einer Polymerlösung konnten sie so mittels UV-Licht miteinander vernetzt werden, woraus mechanisch stabile Endprodukte resultierten. Diese Methode der *akustischen Fabrikation* wird in Zukunft weiter aufmerksam untersucht werden.[16] Durch die Trennung von Partikelanordnung (durch Schall) und Fixierung (durch eine Polymerisationsreaktion) können die einzelnen Prozessschritte separat entwickelt werden, mit dem Ziel, gleichzeitige Fertigung in 3D zu realisieren.

Abschließend untersucht diese Arbeit zwei weitere Ansätze für die akustische Partikelmanipulation.[1] Zuerst wurden Partikel und Tröpfchen in Luft schwebend gefangen. Durch Einsatz eines Reflexionshologramms konnte dabei die übliche Achsensymmetrie eines ein-dimensionalen Resonators aufgelöst und die Partikel paarweise nebeneinander gefangen werden. Der zweite Ansatz nutzt das Hologramm, um Partikel auf einer Wasseroberfläche entlang vorgegebener Routen zu bewegen. Diese Routen werden durch das projizierte Schalldruckbild bestimmt, während der Gradient der Phase einen kontinuierlichen Vortrieb erzeugt.

Contents

Preface

General Introduction

Sound—the propagation of compression or shear waves through matter—is commonly associated with the audible sensation of hearing. However, it is a fundamental feature in mechanics and the basic parameters of sound waves, e.g. amplitude and frequency, span many orders of magnitude. The lower hearing threshold for humans is estimated around 20 μPa pressure, yet people commonly attend music concerts with sound pressure levels at 120 dB of that value. The frequency response of the human ear, too, ranges over three orders of magnitude from 20 Hz to about 20 kHz,[17†]. where values below are termed infrasound and those above *ultrasound*, respectively. Many animals exceed this range, some bats and whales are even sensitive up to 200 kHz.[18] Not only do they use sound for communication among their peers but, more importantly, for navigation and detection of prey. This works through echo-location where the animal sends out a loud but short pulse of sound and then analyses the echos it hears. Those echos contain a wealth of information in their distortion, attenuation, frequency shift and relative phase difference between the animal's two spatially separated ears.[19]

Ultrasonic imaging in medicine and industry uses the same principles of echo-location to map the insides of heterogeneous bodies. Due to the low attenuation of sound waves in biological tissues the operating frequencies could be pushed well into the Megahertz regime, corresponding to

[†]Throughout this text citations are marked by superscript numbers. See List of References on pp. 101.

wavelengths down to hundreds of micrometers. Shorter wavelengths lead to better spatial resolution since the probing wave is influenced by smaller scatterers, which are the sources for the back echos.[10]

It has long been known that sound waves exert forces on matter and exploiting those effects also merits technical advances. 1980 saw the first use of acoustic shockwaves to destroy kidney stones in a human subject.[20] In current medical research, high-intensity focused ultrasound (HIFU) is used to initiate necrosis (via thermal degradation of proteins) of tissue without direct contact, e.g. in the treatment of cancerous tumors that are hard to reach.[21] Ultrasound elastography is a promising extension to ultrasonic imaging, where the sound waves (or other means) define a strain or shear stress and the tissue's elastic response is recorded by the same imaging tool.[22] That reveals the local stiffness variations of tissue inside the body, which is an indicator for various diseases. The photoacoustic effect, first described by Alexander G. Bell around 1878,[23] is another curious effect linking rapid thermal expansion to sound radiation. It has been well explored and refined over the last decades and several implementations of photoacoustic imaging already entered clinical trials.[24]

Acoustic forces can also be used for particle manipulation. Most notable are standing wave configurations that lead particles to migrate to the nodes or anti-nodes depending on their material's make-up, a motion called *acoustophoresis*.[25] Resonator concepts have been realized at centimeter scale to levitate small objects or droplets in air.[26] Much research effort over the last decades went, however, into ultrasonic micromanipulation in lab-on-a-chip devices, where acoustophoresis provides contact-free, label-free enrichment and separation of micrometer-sized particles or biological cells from liquid flows to enhance medical diagnosis and monitoring.[27]

Motivation

The acoustic hologram, which has been developed as part of this thesis, provides an elegant way to modulate the wavefront of a passing sound wave. As such it enables control over 3D sound fields with several orders of magnitude more degrees of freedom (DOF) than has been possible with conventional techniques. In this regard, it will improve existing ultrasonic technologies and possibly enable novel applications of sound waves. This thesis is most concerned with acoustic particle manipulation, where several different approaches have been implemented and investigated. Acoustic fields generated by a hologram might be a route towards instant, one-shot fabrication, where the product shape is defined through the holographic acoustic field and fixation results via another chemical or physical input.

3D-printing provides a cost-effective solution to fabricate the acoustic hologram. Thereafter it is a simple yet powerful tool to investigate advanced manipulation using sound waves. It should be noted that all the concepts investigated as part of this work, from computation of holograms to the assembly and fabrication of composite structures, are not limited to static printed elements. They can be implemented in a similar manner on dynamic phased array transducers (PATs), if those reach sufficient element counts.

Outline

The thesis is structured as follows

- **Chapter 1** reviews the basic equations describing sound waves and their propagation through space. The Rayleigh integral is introduced and it is shown how the angular spectrum method is a more efficient method to compute diffraction. This is followed by a short summary of the different methods to create acoustic fields in the ultrasound range.

- **Chapter 2** introduces the concept of holography, starting with the historical perspective. Then different approaches to compute holograms for a particular field are described. Finally, this chapter covers physical implementations of acoustic holograms, e.g. as a 3D-printed phase plate.

- **Chapter 3** provides experimental validation for the basic concept of acoustic holograms. Which fields can be encoded in a single hologram and how does it affect the image quality? Different use cases are presented, for amplitude and phase control as well as multi-image reconstruction. Each one requires modifications to the basic optimization algorithm, which are provided. Hydrophone measurements validate each case against the simulations.

- **Chapter 4**: The standard method to map ultrasound fields is point-by-point scanning using a hydrophone. This is inherently slow and does not scale well to 3D sound fields. In this chapter, an alternative measurement concept is presented that detects sound intensity via the heating effect it has on an absorbing membrane. A thermal camera continuously records its surface temperature and permits the mapping of entire planes. Translating the membrane through the region of interest results in much faster scanning times for volume acquisitions.

- **Chapter 5**: This chapter introduces the acoustic radiation force, starting with its computation for simple cases, which form the basis of particle manipulation. The intricate sound fields generated by acoustic holograms can be used to assemble microparticles into arbitrary shapes. Finally, a fabrication technique is demonstrated that uses acoustic assembly of particles to form a product, which is then fixed via a crosslinking reaction.

- **Chapter 6** discusses alternative acoustic manipulation concepts. The first part focuses on levitation of millimeter-sized particles or liquid droplets in air using acoustic reflection holograms. The second part presents seemingly dynamic and perpetual propulsion of objects surfing along fixed trajectories on the water surface, which are defined and projected through the acoustic holgram.

List of Publications

The results presented in this thesis have in part been published. In particular, excerpts and figures throughout this thesis were taken from the following journal articles:

- K. Melde, A. G. Mark, T. Qiu, and P. Fischer. "Holograms for acoustics". *Nature* **537** (7621): 518–22 (2016). DOI: 10.1038/nature19755

- K. MELDE, T. QIU, A. MARK, and P. FISCHER. "Akustische Hologramme steuern Partikel". *Physik in unserer Zeit* **48** (1): 9–10 (2017). DOI: 10.1002/piuz.201790014

- K. MELDE, E. CHOI, Z. WU, S. PALAGI, T. QIU, and P. FISCHER. "Acoustic Fabrication via the Assembly and Fusion of Particles". *Advanced Materials* **30** (3): 1704507 (2018). DOI: 10.1002/adma.201704507

- K. MELDE, T. QIU, and P. FISCHER. "Fast spatial scanning of 3D ultrasound fields via thermography". *Applied Physics Letters* **113** (13): 133503 (2018). DOI: 10.1063/1.5046834

The author has been involved in other research projects during that time and coauthored the following publications, which are not covered in this thesis.

- S. PALAGI, A. G. MARK, S. Y. REIGH, K. MELDE, T. QIU, H. ZENG, C. PARMEGGIANI, D. MARTELLA, A. SANCHEZ-CASTILLO, N. KAPERNAUM, F. GIESSELMANN, D. S. WIERSMA, E. LAUGA, and P. FISCHER. "Structured light enables biomimetic swimming and versatile locomotion of photoresponsive soft microrobots". *Nature Materials* **15** (6): 647–53 (2016). DOI: 10.1038/nmat4569

- T. QIU, S. PALAGI, A. G. MARK, K. MELDE, F. ADAMS, and P. FISCHER. "Wireless actuation with functional acoustic surfaces". *Applied Physics Letters* **109** (19): 191602 (2016). DOI: 10.1063/1.4967194

- T. QIU, S. PALAGI, A. G. MARK, K. MELDE, F. ADAMS, and P. FISCHER. "Active Acoustic Surfaces Enable the Propulsion of a Wireless Robot". *Advanced Materials Interfaces* **4** (21): 1700933 (2017). DOI: 10.1002/admi.201700933

- T. QIU, F. ADAMS, S. PALAGI, K. MELDE, A. MARK, U. WETTERAUER, A. MIERNIK, and P. FISCHER. "Wireless Acoustic-Surface Actuators for Miniaturized Endoscopes". *ACS Applied Materials & Interfaces* **9** (49): 42536–43 (2017). DOI: 10.1021/acsami.7b12755

- F. ADAMS, T. QIU, A. MARK, K. MELDE, S. PALAGI, A. MIERNIK, and P. FISCHER. "Wireless micro-robots for endoscopic applications in urology". *European Urology Supplements* **16** (3): e1914 (2017). DOI: 10.1016/s1569-9056(17)31148-x

List of Acronyms

SW	space-bandwidth product
ARF	acoustic radiation force
AS	angular spectrum
CMOS	complementary metal-oxide-semiconductor
CMUT	capacitive micromachined ultrasonic transducer
CT	computed tomography
DMPA	2,2-dimethoxy-2-phenylacetophenone
DOF	degrees of freedom
FFT	fast Fourier transform
FT	Fourier transform
GS	Gerchberg-Saxton algorithm
HIFU	high-intensity focused ultrasound
IASA	iterative angular spectrum approach
IIR	isobutylene isoprene rubber
IO	input-output algorithm
IR	infrared radiation
LASER	light amplification by stimulated emission of radiation
LC	liquid crystal
MEMS	micro-electro-mechanical system
NA	numerical aperture
NEP	noise equivalent pressure

PAT	phased array transducer
PDMS	polydimethylsiloxane
PEG-DMA	polyethylene glycol dimethacrylate
PMMA	polymethyl methacrylate
PS	polystyrene
PVDF	polyvinylidene fluoride
PZT	lead zirconate titanate
ROI	region of interest
SEM	scanning electron microscope
SLM	spatial light modulator
TEM	transmission electron microscope
UV	ultraviolet

Chapter 1

Introduction to Acoustics

This chapter reviews the fundamental equations necessary to describe acoustic fields and to understand holographic image reconstruction via the diffraction of waves. The following sections contain excerpts from the article "Holograms for acoustics".[1]

1.1 The Wave Equation for Acoustics

Acoustic waves in a gaseous or liquid medium are characterized by oscillations of the elemental particles (i.e. molecules) along a propagation direction. The cycles of compression and rarefaction are accompanied by local changes in pressure and density. The field variables to describe the phenomenon are the sound pressure $p(\mathbf{r}, t)$, the particle velocity $\mathbf{u}(\mathbf{r}, t)$ and the density fluctuation $\rho(\mathbf{r}, t)$ for a position $\mathbf{r}$ at time t.

The relations between the individual field variables can be described by the two Euler equations for flow of an ideal, non-viscous and incompressible fluid.[33] The *conservation of momentum* is described by

$$\left(\rho + \rho_{\mathrm{m}}\right)\frac{\mathrm{d}\left(\mathbf{u} + \mathbf{u}_{\mathrm{m}}\right)}{\mathrm{d}t} + \nabla\left(p + p_{\mathrm{m}}\right) = \mathbf{F}, \tag{1.1}$$

where **F** is an external body force density. The *conservation of mass*, on the other hand, yields

$$\frac{\partial\left(\rho+\rho_{\mathrm{m}}\right)}{\partial t}+\nabla\cdot\left(\left(\rho+\rho_{\mathrm{m}}\right)(\mathbf{u}+\mathbf{u}_{\mathrm{m}})\right)=0. \tag{1.2}$$

In both equations variables with subscript m are the non-acoustic values referring to the medium at rest. Those are further assumed to be constant or zero, e.g. the atmospheric pressure p_{m}, the bulk density ρ_{m} or a bulk flow velocity $\mathbf{u}_{\mathrm{m}}=0$.

For the relatively small amplitudes arising in linear acoustics these non-linear equations can be greatly simplified. Then, with $\mathbf{F} = 0$, the fundamental equations describing a linear acoustic field are[33]

$$\rho_{\mathrm{m}}\frac{\partial\mathbf{u}}{\partial t}+\nabla p=0, \tag{1.3a}$$

$$\frac{\partial\rho}{\partial t}+\rho_{\mathrm{m}}\nabla\cdot\mathbf{u}=0, \tag{1.3b}$$

and further

$$p=c^2\rho. \tag{1.3c}$$

Here, c is the speed of sound, i.e. the propagation velocity of a mechanical disturbance in the medium. Equation (1.3c) and Eq. (1.3a) can be inserted into the partial time-derivate of Eq. (1.3b) to yield the homogeneous *wave equation* for the sound pressure

$$\nabla^2 p-\frac{1}{c^2}\frac{\partial^2 p}{\partial t^2}=0, \tag{1.4}$$

and a similar procedure for the particle velocity results in

$$\nabla^2\mathbf{u}-\frac{1}{c^2}\frac{\partial^2\mathbf{u}}{\partial t^2}=0. \tag{1.5}$$

The wavelength $\lambda = c/f$ is the distance between repeating portions of the wave, which oscillates in time with frequency f. In other words λ is the spatial period of the wave. For wavelengths much longer than the mean free path of the molecules in the medium, which is mostly the case for acoustics, the compression cycles proceed adiabatically. In gases this leads to[33]

$$c = \sqrt{\gamma \frac{\mathrm{R}T}{\mu}}, \tag{1.6}$$

where $\mathrm{R} = 8.314\,\mathrm{W\,s\,mol^{-1}\,K^{-1}}$ is the gas constant, γ the heat capacity ratio, T the temperature, and μ the molecular weight. In liquids, on the other hand, the speed of sound depends on the bulk modulus K and the density,

$$c = \sqrt{\frac{K}{\rho_\mathrm{m}}}. \tag{1.7}$$

One of the underlying assumptions to simplify the Euler equations was a flow free of vorticity ($\nabla \times \mathbf{u} = 0$). This also means that $\mathbf{u}$ can be expressed through a *velocity potential* Ψ,[34] so that

$$\mathbf{u} = -\nabla\Psi \tag{1.8a}$$

and

$$p = \rho_\mathrm{m} \frac{\partial \Psi}{\partial t}. \tag{1.8b}$$

Ψ consequently obeys the same wave equation. The velocity potential is a construct often used in text books to determine solutions to a problem, e.g. for scattering of a wave by an object.[34,35] Subsequently, the other field quantities can be calculated from it.

The acoustic field variables in this work are expressed using complex notation, such as

$$\Psi(\mathbf{r},t) = |\Psi|\,\mathrm{e}^{\mathrm{i}\phi(\mathbf{r},t)} = |\Psi|\,\mathrm{e}^{\mathrm{i}(\mathbf{k}\cdot\mathbf{r}-\omega t)}, \tag{1.9}$$

where $\mathrm{i} = \sqrt{-1}$ and $\phi(\mathbf{r},t)$ is called the *phase*. There, $k = \|\mathbf{k}\| = \omega/c$ is the wavenumber in the medium, and $\mathbf{k} = (k_x, k_y, k_z)$ the wave vector, $\omega = 2\pi f$ is the angular frequency, and f the frequency of the wave. The vertical bars around $|\Psi|$ denote the amplitude of a complex variable. The form in Eq. (1.9) presents a solution to the wave equation and the instantaneous field quantities can be obtained by taking the real part of the expression.

Another useful parameter is the impedance Z, which is a property of the traversed medium and relates sound pressure to particle velocity. It is defined as

$$Z = \frac{p}{\|\mathbf{u}\|}. \tag{1.10}$$

Evidently, Z is a complex valued function and its form depends on the problem at hand. As a material property, however, the relation $Z_\mathrm{m} = \rho_\mathrm{m} c$ is commonly referred to, which corresponds to the particular case of a plane wave in an infinite bulk of material. The impedance determines the behavior of waves at the interface between different media. Acoustic waves couple more efficiently into a second medium of the same impedance, which is called an impedance match. On the other hand a mismatch leads to partial or even full reflection of the wave depending on the ratio of the impedances at the interface.

The particle velocity $\mathbf{u}$ is a vector and not trivial to measure experimentally. Using Equation (1.3a), however, it can be determined from the spatial pressure distribution,

$$\mathbf{u} = \frac{\nabla p}{\mathrm{i}\rho_\mathrm{m}\omega}. \tag{1.11}$$

In the case of a plane wave this simplifies further to a scalar $|u| = |p|/Z_\mathrm{m}$.

The acoustic power passing a surface Ω can be calculated as

$$\mathbf{P} = \iint_\Omega \mathbf{I} \cdot \mathbf{n} \, \mathrm{d}A, \tag{1.12}$$

where $\mathbf{n}$ is the normal vector on the area element $\mathrm{d}A$ and the acoustic intensity $\mathbf{I}$ is defined as

$$\mathbf{I} = p\mathbf{u}. \tag{1.13}$$

The acoustic intensity is time-varying according to the field variables, but it has a constant component in the direction of wave propagation. This is the characteristic intensity[35]

$$I_\mathrm{c} = \frac{1}{2} \frac{|p|^2}{\rho_\mathrm{m} c}, \tag{1.14}$$

with $|p|^2 = pp^*$, which can more easily be obtained from sound pressure measurements. The asterisk on p^* denotes a complex conjugate.

1.2 Propagation and Diffraction

Wave propagation and diffraction are essential to understand holography. Given the acoustic field quantities on a surface, how does the field evolve in the volume behind it?

1.2.1 The Diffraction Integral

An intuitive approach divides a large source into many point sources and determines the influence of each source point onto an observation point $O(x, y, z)$. The superposition principle of waves demands that the solution at O is equal to the sum of all point sources' contributions. Reducing the spacing between sources eventually leads to an integral expression, called the Rayleigh integral.[34]

The field radiating from a flat vibrating surface is a typical problem encountered in acoustics. In the literature this is called a piston source mounted on an infinite planar baffle. The baffle is a rigid wall and as such enforces a boundary condition of zero normal particle velocity ($\mathbf{u}\cdot\mathbf{e}_z = 0$). In this thesis all travelling waves propagate by convention along the z-direction. The plane containing source and baffle spans (x, y) at $z = 0$. Consequently, this model only propagates sound into the positive half space $z \geq 0$. The piston source has a uniform time-varying surface velocity $u_0 = |u_0|\cos(\omega t)$, which specifies the boundary condition for particle velocity across the source's extent.[36]

Given the velocity input, the Rayleigh integral is[37]

$$p(x, y, z) = -\frac{\mathrm{i}k\rho_\mathrm{m}c_\mathrm{m}}{2\pi}\iint_{-\infty}^{\infty} u_0(x_0, y_0, 0)\frac{\mathrm{e}^{\mathrm{i}kR}}{R}\,\mathrm{d}x_0\,\mathrm{d}y_0, \tag{1.15}$$

where $p(x, y, z)$ is the sound pressure amplitude at the observation point, (x_0, y_0) are the coordinates in the source plane at $z = 0$ and R is the distance between source and observation points $R^2 = (x - x_0)^2 + (y - y_0)^2 + z^2$.

Alternatively, the sound pressure can be given about a so-called pressure release surface. Physically, this boundary condition better describes an open aperture rather than a vibrating surface. However, in linear acoustics the field variables can be calculated from each other, which renders the two approaches interchangeable. The integral for a pressure input is[37]

$$p(x,y,z)=\frac{z}{2\pi}\iint_{-\infty}^{\infty}p(x_0,y_0,0)\left(\frac{1}{R}-\mathrm{i}k\right)\frac{\mathrm{e}^{\mathrm{i}kR}}{R^2}\,\mathrm{d}x_0\,\mathrm{d}y_0. \tag{1.16}$$

If the field in the entire observation plane is sought, the Rayleigh integral has to be evaluated for each sampling point seperately. It is apparent how field calculations using the Rayleigh integral are computationally intensive.

Faster alternatives or approximations exist for certain cases, which can be identified using the dimensionless *Fresnel number*[38]

$$F=\frac{L^2}{4z\lambda}, \tag{1.17}$$

where L is a characteristic dimension of an aperture, e.g. its diameter, and λ is the wavelength. For $F \ll 1$ the distance in z-direction is much larger than the spatial extent of the source and thus $R \approx z$. This region marks the *far-field* where the Fraunhofer approximation can be applied. The diffraction integral in this case can be replaced by a single Fourier transform (FT).[5]

In ultrasound acoustics, however, typical working distances are of the same order as the size of the source. In the next chapter it will be shown how the information content of holograms benefits from a large ratio of source diameter to wavelength. The Fresnel numbers encountered in this work range from about 3 (6.1 Levitation in Air) to 28 (everywhere else).

1.2.2 The Angular Spectrum Method

In the *near-field*, where the distance to the image is comparable to the source aperture and the lateral dimensions are large relative to the wave length, the Fresnel number $F \gg 1$. Then R varies considerably across all sampling points and it is crucial to keep the exact R-term in Eq. (1.16) so that field contributions of different source points properly add up.

A very effective method to compute wave propagation in this region is to express the integrals in the frequency space. An FT decomposes the source field into its spatial frequencies, called the angular spectrum (AS) of the wave, where each frequency can be thought of as representing an elementary plane wave cutting the source plane at a specific angle. The Rayleigh integral can be interpreted as a convolution of two functions, the source field and a propagator, which can be carried out more efficiently as a multiplication in the frequency space. Computing wave propagation using the AS method therefore yields the exact phase and has relatively low computational cost.[37]

The acoustic pressure wave is expressed as

$$\tilde{p}(x,y,z) = p(x,y,z)e^{\mathrm{i}\phi(x,y,z)} \tag{1.18}$$

where the explicit time dependence has been dropped, and where $p(x,y,z)$ and $\phi(x,y,z)$ are the amplitude and phase function maps, respectively. The AS of this wave in a plane at constant z is obtained by the FT

$$P(k_x,k_y,z) = \iint_{-\infty}^{\infty} \tilde{p}(x,y,z)e^{-\mathrm{i}(k_x x + k_y y)}\mathrm{d}x\mathrm{d}y, \tag{1.19}$$

where k_x, k_y are the spatial frequencies within the plane. The hologram plane is at $z = 0$. Then, once $P(k_x,k_y,0)$ is known, the AS at any plane downstream can be calculated by multiplying the AS in the hologram plane with a propagator function H, so that

$$P(k_x, k_y, z) = P(k_x, k_y, 0)H(k_x, k_y, z), \tag{1.20a}$$

$$H(k_x, k_y, z) = e^{\mathrm{i}z\sqrt{k^2-k_x^2-k_y^2}}. \tag{1.20b}$$

This corresponds to the convolution, where each elementary plane wave is phase-shifted according to its angle of incidence and the propagation distance z. The real-space pressure field in the plane at z can be obtained via the inverse FT

$$\tilde{p}(x, y, z) = \frac{1}{4\pi^2} \iint_{-\infty}^{\infty} P(k_x, k_z, z)\mathrm{d}k_x \mathrm{d}k_y. \tag{1.21}$$

For infinite planes the AS has been shown to exactly represent the scalar diffraction integral.[5] Finite size apertures and discrete sampling points, however, restrict the bandwidth of the AS and thereby introduce errors and artifacts. This can be mitigated by chosing proper domain sizes and sampling frequencies. For a single propagation a computational domain size of

$$L_\mathbf{C} \geq 3 \times \max\{L_\mathbf{H}, L_\mathbf{I}\} \tag{1.22}$$

is sufficient, where $L_\mathbf{H}$, $L_\mathbf{I}$ refer to the largest dimension of the hologram or image region, respectively. According to the Nyquist-Shannon sampling theorem the distance between sampling points Δ should be smaller than or equal to half the spatial period of the wave, i.e. $\Delta \leq \lambda/2$.

The back-propagation can be performed similarly by

$$P\left(k_x, k_y, 0\right) = P\left(k_x, k_y, z\right) H^{-1}\left(k_x, k_y, -z\right), \tag{1.23}$$

where the inverse propagator H^{-1} has been modified to attenuate evanescent waves, which would otherwise rise exponentially, greatly amplifying any numerical error.[37]

$$H^{-1}\left(k_x,k_y,-z\right)=\begin{cases} e^{-\mathrm{i}z\sqrt{k^2-k_x^2-k_y^2}} & ,\ k_x^2+k_y^2\leq k^2 \\ e^{-z\sqrt{k^2-k_x^2-k_y^2}} & ,\ k_x^2+k_y^2>k^2 \end{cases}. \tag{1.24}$$

1.3 Generation of Ultrasound Waves

Sound waves can be generated by any interface that moves periodically and thus starts a disturbance in a medium that propagates further at the speed of sound. There are different electro-mechanical, but also optical methods to generate sound waves. This sections highlights selected approaches that are relevant for ultrasound.

A piston source vibrating with the normal velocity $u_0 = |u_0|\cos\omega t$ causes the molecules at the interface to follow its motion without slip. The surface displacement is found via integration, $s_0 = |u_0|/\omega\cdot\sin\omega t$, and the sound pressure follows from the acoustic impedance of the medium $|p_0| = |u_0|Z_\mathrm{m}$. For example, a surface oscillation into water with displacement amplitude of 1 nm at 2 MHz causes a sound pressure amplitude of 18.6 kPa.

1.3.1 The Piezoelectric Effect

Certain dielectric materials, particularly crystals or ordered materials, can accumulate electrical charge in proportion to applied strain, when the medium lacks inversion symmetry. Strain due to deformation induces charge separation and the resulting dipoles add up across the material to a net electric field. The interplay of mechanical and electrical variables can be described by two equations,[39]

$$D = \epsilon^{\sigma} E + d'\sigma, \tag{1.25a}$$

$$S = dE + s^{E}\sigma, \tag{1.25b}$$

where D is the electric displacement field, E the electric field, and σ, S are mechanical stress and strain, respectively. They are interconnected through the permittivity ϵ^{σ} at constant stress, the direct and inverse piezoelectric coefficients d, d', and the compliance s^{E} under short circuit condition ($E = 0$). The process is reversible, so that an applied electrical field causes the crystal to deform. This is called the *inverse* piezoelectric effect and it is the basis for most ultrasound transducers.

Examples for piezoelectric materials are the naturally occuring quartz (SiO_2) or the synthetic ceramic lead zirconate titanate (PZT), which is currently the most widely used material for ultrasound transducers.[39] A typical value for the longitudinal piezoelectric coefficient in commercial PZT is $d_{33} = 265 \times 10^{-12}\,\mathrm{C\,N^{-1}}$ (PIC181, PI Ceramic, Germany). By convention the direction 3 is along the axis of polarization, so that the subscript 33 indicates strain along the same dimension. Using this value an unclamped crystal (i.e. $\sigma = 0$) of 1 mm thickness requires a potential difference of 4 V between its sides to expand by 1 nm. It is evident how an applied sinusoidal voltage can be used to radiate a sound pressure wave into an adjacent medium.[40]

However performant it is, the lead content in PZT poses environmental and health risks, which in turn sparked research efforts to find a more benign material replacement. There are polymers that possess piezoelectric properties, although with much lower magnitude than ceramics. One notable example is electrically poled polyvinylidene fluoride (PVDF). Its low acoustic impedance and processibility make PVDF a good choice for large bandwidth ultrasound sensors, such as hydrophones.

The design strategies for most ultrasound transducers employed in imaging and non-destructive testing are similar.[41] A flat disc of piezoceramic material is in contact with a matching layer (thickness $\lambda/4$) that improves wave coupling into the medium. The electrodes are usually deposited across the large faces and the transducer is said to operate in *thickness mode.* Transducers for imaging require a wide frequency bandwidth for pulse echo measurements. This is accomplished through an attenuating backing layer, which reduces the quality factor of the resonant acoustic circuit and thereby increases the bandwidth of the transducer.

High-power transducers, on the other hand, benefit from resonant amplification. The most common design is credited to Paul Langevin who invented it around 1918 while working on underwater echo-detection of submarines.[42] The *Langevin transducer* is composed of a piezoelectric disc that is press-clamped between two metallic cylinders. The total length of this assembly is a multiple of $\lambda/2$ to achieve a standing wave upon excitation resulting in large displacement amplitudes at the ends. A reduction of the cross-section further amplifies this amplitude. That mechanical amplifier is called an ultrasonic horn and such constructs are often found in devices for ultrasonic welding and liquid processing but also acoustic levitation in air.

1.3.2 Alternative Mechanisms

In general, ultrasound waves are radiated from any interface vibrating at a high enough frequency and there are alternative methods to using piezoelectric materials. One example is the capacitive micromachined ultrasonic transducer (CMUT). A CMUT cell is a capacitor with two electrode plates separated by a vacuum. An applied voltage leads to electrostatic attraction between the plates, which is resisted by their mechanical stiffness. To generate ultrasound an alternating voltage has to be applied.[43] Advantages of CMUT are a wide achievable bandwith, the compatibility with micro-electro-mechanical system (MEMS) fabrication and complementary metal-oxide-semiconductor (CMOS) processes, which enable direct integration with microelectronic circuitry and no dependency on potentially hazardous ceramics such as PZT. Hence, CMUT are a promising technology for next generation PATs in biomedical imaging.

Thermal expansion can also be exploited for sound generation, by using the photoacoustic effect. For a sound wave to propagate from a rapidly expanding volume element, two constraints have to be met: Firstly, the heating has to be faster than thermal relaxation due to heat conduction (thermal confinement). Secondly, the heated volume has to expand at a faster rate than the speed of sound (stress confinement). In biological tissues these conditions are met by nanosecond laser pulses.[23] This mechanism is best suited to pulsed or short burst excitation rather than continuous operation. Photoacoustic imaging is a promising technique in medical imaging, because it combines optical excitation, which can be very specific due to labeling, with advantages of ultrasound imaging, such as low attenuation and ease of detection and measurement.

1.4 Recording of Ultrasound Fields

The direct piezoelectric effect can be used to sense and measure sound waves. Most ultrasound transducers are designed for pulse-echo measurements and thus employ both the direct and inverse piezoelectric effect. To improve the frequency response transducers made from hard ceramics such as PZT are typically damped by an attenuating layer connected to their backside.

Some alternative materials are better suited to precision measurements. For example, membranes made from polymers such as PVDF show exceptionally good sensitivity since their low acoustic impedance matches better with aqueous media than ceramics do. Reducing the membrane thickness to very thin values (common values are about 10 μm) increases the resonance frequency considerably and thus enables a flat signal response over a larger frequency range.[33]

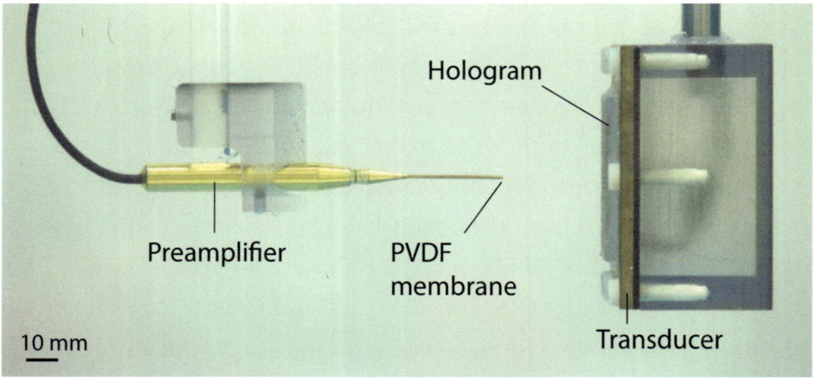

Figure 1.1: Ultrasound field mapping with a needle hydrophone, which consists of a preamplifier body and an interchangeable needle tip hosting the PVDF membrane. The hydrophone is translated across the region of interest by an automatic scanning stage.

The physical presence of the sensor in the sound field causes reflection and diffraction of the wave. This in turn leads to errors due to the interference with the measured field. One solution is the minimization of the sensor size. The corresponding device is called a needle hydrophone and a photograph of a commercial product (Precision Acoustics, UK) is shown in Figure 1.1. The body is a hollow metal needle with its tip covered by a gold-plated PVDF membrane. Needle diameters are available down to several hundred micrometers. This sensor is placed head-on into the sound field providing a precise amplitude measurement at a single point. The relative phase for each sampling point can be obtained by comparison of the signal with a reference wave, e.g. the transducer excitation. Hydrophone scans are used extensively in Chapter 3 to validate the reconstructed holographic pressure fields. Chapter 4, on the other hand, presents an alternative method to quickly acquire volumetric maps of sound pressure.

Chapter 2

Holograms for Acoustics

This chapter introduces the concept of holography as a fundamental feature of all wave fields, starting with the historical perspective and development of basic optical holography. This is followed by a survey of common approaches to compute holograms and their optimization regarding particular fields and transducers. The final section then covers physical implementations of acoustic holograms, for example as a 3D-printed phase plate. This chapter is based on and contains excerpts from the article "Holograms for acoustics".[1]

2.1 Holography

In conventional imaging, for example photography, the *intensity* of an optical wave or electron beam is stored via photochemical or photoelectrical processes in a planar medium. Much spatial information is lost in this projection and different 3D scenes can result in the same recorded perspective. The history of a wave, which contains its propagation through space, can , however, be recovered from a full recording of its wavefront. That wavefront is expressed in the complex phase of the wave.

Holography has been invented by Dennis Gábor in 1948 as a method to improve transmission electron microscopy. At that time it was well-known that much insight about the specimen is hidden in the phase information of the altered probing wave. However, the high frequencies of electromagnetic waves complicate direct phase measurements. Gábor's idea was to use interference of the measurement wave with a second wave—called the reference wave—to access the phase information in the form of interference fringes in the recorded intensity image.[7] In 1971 he was awarded the Nobel Prize in Physics for this work. Gábor aptly named his recordings *holograms,* derived from the greek words ὅλος (holos) and γράμμα (gramma), literally meaning whole recording. In his setup the reference beam was simply the unscattered order of the probing beam and thus its amplitude was much larger than the scattered wave component. Image reconstruction in holography is performed by illuminating the hologram with a reference beam. Using the same on-axis setup as Gábor this results in paraxial reconstruction with two conjugate images and a bright spot of the zeroth order in the center.

For interference to happen the two beams have to be coherent. Several years later, in 1960, the invention of the LASER (light amplification by stimulated emission of radiation) made sources of coherent visible light available and consequently research on holography boomed in the following decade. A major improvement to holography, introduced by Emmett N. Leith and Juris Upatnieks,[44] put object beam and reference beam at an oblique angle. Their modification separates the zeroth order and the conjugate image (also called twin image) from the actual image, thus producing a much clearer result.

The underlying principle, that a wave field can be reconstructed from a full recording of its complex amplitude and phase about a sufficiently large aperture, is true for all wave fields, including acoustic waves. The relatively low frequencies of acoustics, however, make it possible to directly record the complex field quantities. Another distinction to optics are the relaxed coherence criteria. The sound waves of any two independent sources can interfere with each other over large distances, provided their frequencies match.

Classic holograms were used to record *images* of physical objects in a way that includes spatial information. At a later stage the images in place of the object could be reconstructed by illumination with a reference beam. Typically, the goal was to provide a virtual image to an observer or a real but scaled image for some metrology tool.[8]

2.1.1 Classification of Holograms

This poses the question if the recording step is even necessary. Is it possible to create arbitrary fields using holography—potentially in 3D? How does the physical hologram look that encodes this information? The solution takes two steps:

1. Computation of the hologram (i.e. the *wavefronts*) that will diffract upon propagation to form the desired field distribution and

2. the physical implementation in the experimental setting.

In step 1 several assumptions about the system and its dimensions are necessary, including sound frequency and wavelength, image distance and aperture size. Since the linear wave equation is time-symmetric the most trivial solution to find a hologram is to arrange the desired image as a distribution of point sources and virtually propagate the total wave field to the hologram aperture.

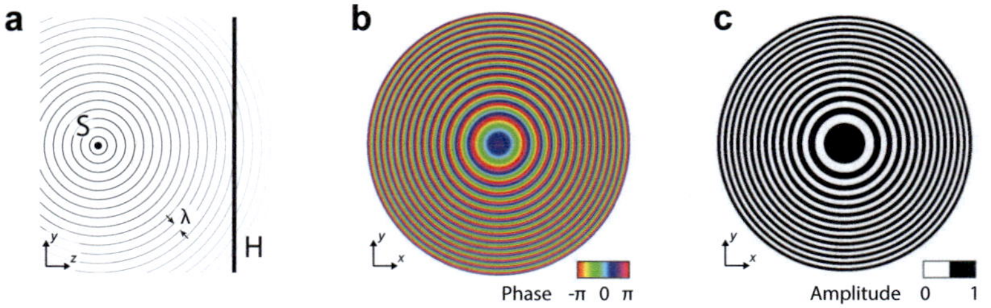

Figure 2.1: Hologram of a point. (**a**) Spherical wave fronts from point source **S** are recorded along a virtual hologram plane **H**. (**b**) Phase hologram and (**c**) binary amplitude hologram for a point image. For both examples the shortest distance from **S** to **H** is 15λ and the aperture diameter of **H** is 50λ.

Figure 2.1(a) visualizes the problem for a single point source **S**, which radiates a spherical wave in all directions.[34]

$$\Psi(r,t) = \frac{1}{\mathrm{i}kr}|\Psi|_{\mathrm{S}}\,\mathrm{e}^{\mathrm{i}(kr-\omega t)}. \tag{2.1}$$

The wave can be recorded anywhere, for example across plane **H**, resulting in a hologram. The recorded amplitude is $\sim 1/R$, where R is the distance between point **S** and a point on **H** and the recorded phase at a time instant is $\phi = kR$. The image of point **S** can be reconstructed from the hologram by sending the recorded wave back in time. Since time cannot be turned back, the same is achieved by taking the complex conjugate of the recorded phase, $\phi = -kR$. This converts the diverging wave to a converging wave focusing towards **S** in this case. Unfortunately, there are several issues with this method: The size and location of **H** limits the information that is recorded to a small section of the original spherical wave and the remaining field will be lost. Further, a converging wave becomes a diverging wave when passing the focal point **S**. Perfect inversion of the original field logically requires a sink at **S**, such as an ideal absorber.[45]

The synthesis of arbitrary fields using digital holography is well established in optics.[8] Virtual computation frees the design from the practicality of an experimental recording. However, the wave still has to be physically reconstructed at some point. This requires methods to modulate wavefronts. In optics those methods include photographic emulsions, diffraction gratings or spatial light modulators based on liquid crystal cells or micromirror devices. Most of these devices only modulate either the phase or the amplitude of a passing wave and thus cannot encode the exact hologram.[5] Using optimization algorithms it is possible to compute approximate holograms that work within the restrictions imposed by the modulation device yet reconstruct to fields of satisfying quality.

Figure 2.1(b) and (c) present two examplary holograms that work by controlling only phase or only amplitude, respectively. Figure 2.1(b) is a *phase hologram*, equal to the inverted recorded phase and limited to a finite circular aperture. An ideal phase hologram does not modulate the amplitude, which is then solely defined by the incident reference beam. Figure 2.1(c) is a *binary amplitude hologram*. Each point across **H** has a fixed pathlength R towards **S** and therefore a fixed phase relation $\phi = kR$. The black rings are chosen so that only points with phases $\phi \in [-\pi/2, \pi/2]$ transmit full amplitude and the other points block the reconstruction wave. Thus only the parts of the wave, which interfere constructively at **S**, are transmitted. This special example is also known as a binary fresnel zone plate.

2.2 Computation of Holograms

In practical systems it is often not possible to control both the amplitude and phase values of the reconstruction wave. Various optimization algorithms have been developed for this reason to compute approximate holograms, which reconstruct to acceptable image qualities. Most of these algorithms have been developed for imaging applications but are applicable to field synthesis as well.[46]

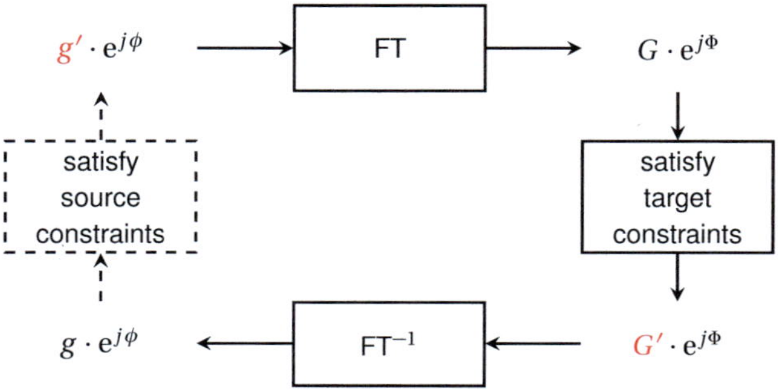

Figure 2.2: Flow diagram of the Gerchberg-Saxton algorithm (GS). Spatial domain (left) and frequency domain (right) are related via Fourier transforms (FTs). The measured intensity patterns (inputs) are highlighted in red. Opening the feedback loop turns it into Fienup's input-output algorithm (IO).

2.2.1 Error-reduction Algorithms

The most influential algorithm bears the name of Ralph Gerchberg and Owen Saxton, who published it in their seminal article about determination of phase from microscope images, specifically the transmission electron microscope (TEM).[47] In such a microscope one can switch between two imaging modes recording distinct but related intensity patterns for the same specimen—its image and the corresponding diffraction pattern. In the Fraunhofer approximation, which is applicable here, the far-field diffraction pattern is equal to the FT of the image and vice versa. This knowledge can be used to improve the signal, because any image feature needs support in the frequency domain.

The Gerchberg-Saxton algorithm (GS) is a very efficient way to correct the image and the diffraction pattern in an iterative cyclic process comparable to a feedback loop. A schematic is shown in Figure 2.2. The recorded intensity patterns g' and G' in the spatial and frequency domain, respectively, are highlighted in red. The algorithm takes the recorded image g' and after a FT compares its complex valued signal in the frequency domain to the recorded diffraction pattern G'. Then only the amplitude is corrected to the measured value, while the complex phase is kept. The result is propagated back to the spatial domain via an inverse FT, where the amplitude is changed to the measured signal g'. Transforming back and forth between the two planes quickly converges to a solution—which is the phase information of the image or, in other words, the phase hologram ϕ for a particular field. The algorithm ends when a sufficiently good image is obtained. The next section discusses metrics to assess reconstruction efficiency. Other common names for the GS are error-reduction algorithm or ping-pong algorithm.

The input-output algorithm (IO) introduced by James R. Fienup opens the feedback loop and thereby allows any input to be used.[46] Assumptions or knowledge about the object's support in the spatial or frequency domain can then be used to improve the image. This change is reflected in Figure 2.2 by removal of the dashed elements. The input g' is then a driving function that does not necessarily have to satisfy the object constraints. This can be useful if only one intensity measurement is available as for example in astronomical observations or the synthesis of arbitrary fields, such as for holographic optical tweezers.[48]

It is still not fully understood why these algorithms perform so well,[49] although the error has been shown to decrease or stagnate for each iteration with GS. For a single intensity measurement the GS has been shown to be identical to the steepest descent gradient method.[46] Since these algorithms are in some form similar to gradient methods they are prone to settle in any local minimum. To prevent that from happening one can add some entropy to the inputs, such as modifying the phase at random points. This has been combined with other optimization strategies like, for example, simulated annealing.[50]

Computational synthesis of three-dimensional fields is still an active field of research, with many open fundamental questions. For example, it is not trivial to find the ideal synthesis for a given 3D target distribution if the actual position, orientation or even number of sources and holograms is left open and sought after.[6]

2.2.2 Iterative Angular Spectrum Approach

In optical holography the image is usually reconstructed in the far-field, where the Fraunhofer approximation is applicable. Hologram and image are then related via the FT. To move the image closer to a finite distance the Fourier-transforming properties of a lens can be used. The FT relation leads to other well-known properties of these Fourier holograms, such as being able to reconstruct the whole image from just a small section of the hologram, albeit with less fidelity.

The typical dimensions in acoustic experiments are much smaller, on the order of hundreds down to tens of wavelengths, and acoustic lenses are not commonly available. At short distances, where the lateral dimensions are of the same order as the distance between hologram and image plane, the Fraunhofer and Fresnel approximations generate large errors. In this case it is necessary to keep the exact phase term during propagation.[5] The angular spectrum (AS) method is valid in the near-field and consequently used in propagation of acoustic fields. By using the fast

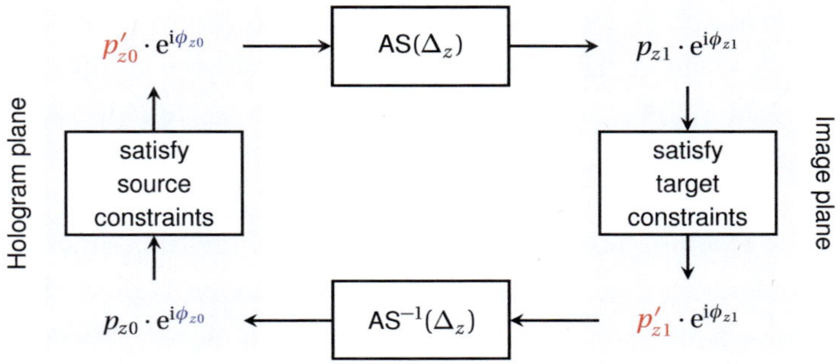

Figure 2.3: Flow diagram of the iterative angular spectrum approach (IASA). The field is propagated from hologram plane (left side) to one or several image planes (right side) using the angular spectrum (AS) method. Then in each plane the desired amplitude distributions are inserted as boundary conditions. After a number of iterations ϕ_{z0} eventually contains the solution for the phase hologram.

Fourier transform (FFT) it performs much better than numerically solving the diffraction integral. As discussed in Section 1.2 one projection step of the AS uses two FTs and one multiplication with a propagator function. The iterative angular spectrum approach (IASA) is a modification of the GS algorithm, where the FT has been replaced by AS propagation.

Figure 2.3 shows the updated flow diagram for the IASA, where $p_{z1} \cdot e^{j\phi_{z1}}$ describes the field in the image plane $\mathbf{I}$ over (x, y) at a constant z_1. The target region $\mathbf{T}$ is defined as the points with non-zero amplitude and a subset of the image region $\mathbf{I}$. The field in the hologram plane is in $p'_{z0} \cdot e^{j\phi_{z0}}$. One completed loop in this diagram equals one iteration of IASA, which can be divided into four steps:

1. The field from the hologram plane is propagated via the AS method to one or several image planes at distances $\Delta_{z,i} = z_i - z_0$.

2. Constraints in the image plane are applied, which typically is the desired amplitude distribution. In this case the complex phase at each point is kept and only the amplitudes are replaced by the desired values p'_{z1}.

3. The resulting field is back-propagated via the AS method to the hologram plane. If there are several image planes, they are propagated separately and then merged into one hologram by complex addition.

4. The source constraints p'_{z0} are applied in the hologram plane. Those are defined by the reconstructing beam, i.e. the transducer output, which is to be modified by the hologram.

Wave propagation can be seen as a low pass filter, where the propagation distance limits the highest spatial frequency that contributes to an image. The computational domain is limited in space. When using the AS method for propagation, however, the highest spatial frequencies will not be lost at the domain boundaries. Instead they are reflected back into the domain in an effect called *aliasing*. To address this problem the domain's extent has to be sufficiently large to allow all bandwidth-limited spectral components emanating from the hologram or image aperture to propagate without aliasing. According to Eq. (1.22) a sufficient domain size is about three times the largest dimension L_{H}, L_{I} of either hologram or image region, respectively. The field outside the hologram aperture is further set to zero during optimization. Additionally, it is benefitial to limit the integration region of the inverse Fourier transform to exclude high frequency components that cannot contribute to image formation. This reduces the error caused by successive propagation steps and multiple iterations (as required by the IASA). A simple geometrical estimate yields[51]

$$k_{x,y} \leq \frac{\pi(L_{\mathbf{H}} + L_{\mathbf{I}})}{\lambda\sqrt{\frac{1}{4}(L_{\mathbf{H}} + L_{\mathbf{I}})^2 + \Delta_z^2}}, \tag{2.2}$$

where λ is the wavelength in the medium. The algorithm ends when a sufficiently good image is obtained. The final phase hologram is contained in ϕ_{z0}, which is optimized for an amplitude distribution p'_{z0} at the input. Metrics to assess reconstruction quality and efficiency will be discussed in the next section.

2.3 Error Metrics for Image Reconstruction

There are different ways to express the quality of holographic reconstruction. As a measure of overall reconstruction efficiency η it is useful to relate the acoustic power in the target region **T** to the total acoustic power in the whole image plane **I**. The region **T** is the set of image points that was designed to have a non-zero amplitude according to the target image in **I**. For simplicity the wave is assumed to pass the observation plane in normal direction and that pressure and particle velocity are in phase. Those assumptions correspond to a plane travelling wave. Then the acoustic intensity can be calculated from the sampled pressure amplitudes as $I_{z,i} = p_{z,i}^2/2\rho c_{\mathrm{m}}$. The power can in turn be obtained by summation of the intensity over all sampling points

$$P_z = \Delta x \Delta y \sum_{i \in \mathbf{I}} I_{z,i}, \tag{2.3}$$

where the $\Delta x, \Delta y$ are sampling distances along x and y coordinates in the observation plane, respectively. In computation and experiment a sampling distance $< \lambda/2$ is recommended. The reconstruction efficiency then follows as

$$\eta = \frac{P_{z,\mathbf{T}}}{P_z} = \frac{\sum_{i\in\mathbf{T}} p_{z,i}^2}{\sum_{i\in\mathbf{I}} p_{z,i}^2}, \tag{2.4}$$

which is a comparable metric to the diffraction efficiency in optics.[50]

Optimization requires a gauge of improvement in the form of a cost function C that has to be minimized. In the case of holography C can intuitively be expressed as the squared error of the field amplitudes[50]

$$C = \sum_{i\in\mathbf{I}} \left(p_{z,i} - p'_{z,i}\right)^2, \tag{2.5}$$

where p'_i is the target amplitude at the ith pixel and the summation extends over all points in the image plane $\mathbf{I}$. For $C \to 0$ the computed field approaches the target image.

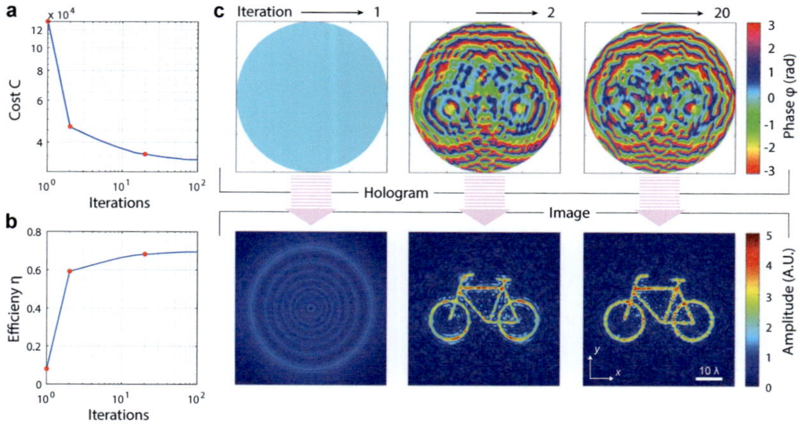

Figure 2.4: Hologram computation via iterative angular spectrum approach (IASA). (**a**) Cost function to be minimized. (**b**) Efficiency of simulated reconstruction. (**c**) Snapshots for iterations 1, 2 and 20, showing the phase map in the hologram plane (top row) and the resulting amplitude distribution in the image plane (bottom row). Scale bar, ten wavelengths λ.

Figure 2.4 illustrates the hologram computation via IASA for a simple image. The cost function C in panel (a) converges quickly to a (local) optimimum and the reconstruction efficiency η in panel (b) shows similar behaviour. For three chosen instants during the optimization the initial phase maps in the hologram plane (top row) and associated amplitude maps in the image plane (bottom row) are shown. Iteration 1 propagates the plain transducer output resulting in the characteristic diffraction pattern of a piston source. In the second iteration, the hologram already contains the phases of back-projected target pixels and its reconstruction shows an image with η of almost 60 %. Additional cycles of IASA slightly improve the result. The holograms in this work were typically computed using between 20–50 iterations.

Being limited by a finite and flat aperture at a fixed distance to the image restricts the possible fields that a hologram can encode. Often the perfect reproduction is impossible but approximations lead to acceptable outcomes. The errors typically impart the homogeneity of the amplitude distribution, or show as pointwise intensity concentration (speckle) or ghost features repeating periodically. The optimization goal always has to involve on the application whether it is to maximize power or to provide a more uniform image.

2.4 Implementation of Holograms for Acoustics

The algorithms in Section 2.2 produce a phase hologram for a desired target image. This section explores several techniques to generate the required acoustic wavefronts as a prerequisite to reconstruct the image in an experimental setting.

First, this task requires thinking about the apparent complexity of a wavefront. This can be interpreted as the information content of a wave, which can be quantified by the scalar space-bandwidth product (*SW*).[52] Because the *SW* is invariant, a transmission hologram's *SW* limits the intricacy of the acoustic field in the workspace downstream from it, down to the diffraction limit. In a digital hologram of finite size, consisting of $N \times N$ discrete pixels, the *SW* is given by $SW = N^2$. Thus the information content of the acoustic field depends directly on the number of pixels that can be independently controlled in the hologram that generates it. Qualitatively, holograms with higher *SW* lead to reconstructed fields with larger spatial extent and finer detail. The *SW* is related to the DOF of a projection or imaging system.[53]

The space-bandwidth product of a square-shaped hologram **H** is $SW_\mathbf{H} = N \times N = (L_\mathbf{H}/\Delta)^2$, where $L_\mathbf{H}$ and Δ are the lateral sizes of the hologram and the pitch between neighbouring pixels, respectively, and N is the number of pixels in each dimension. A pixel is a small area element that individually controls a portion of the transmitted or reflected wave.

Regarding the image **I**, its space bandwidth product is $SW_\mathbf{I} = L_\mathbf{I}^2 B_\mathbf{I}^2$, where $B_\mathbf{I}$ is the spatial bandwidth of the image. Free space propagation cannot increase the *SW*, so $SW_\mathbf{I} \leq SW_\mathbf{H}$. But the upper bound of the image's spatial bandwidth is set by the diffraction limit, $\sim \lambda/2$, so that $B_\mathbf{I} \leq 2/\lambda$, and $SW_\mathbf{I} \leq 4(L_\mathbf{I}/\lambda)^2$. Diffraction-limited performance is achieved when $SW_\mathbf{H} = SW_\mathbf{I}$, which reduces to $\Delta = \lambda/2$ as the optimal sampling distance.[1]

In short, the hologram has to match the complexity of the stored image and the modulation device, which encodes the hologram, needs to provide a minimum number of pixels to accomplish this task. A pixel spacing of $\lambda/2$ is sufficient to reproduce the wave front. The following sections investigate different modulation devices for ultrasound.

2.4.1 Phased Array Transducer (PAT)

Phased array transducers (PATs) consist of smaller transducer elements that each radiate a sound wave and the superposition of all these small contributions is a more complex wave.[11] Figure 2.5(a) shows the operating principle of PATs, where each element receives an electronic signal with a specific phase offset. This in turn leads to a phase lag of the radiated wave and it can be imagined how the right addressing scheme creates any abritrary wave front. Typical element counts of commercial PAT products range from 8 to about 1000 and they are configured in linear, rectangular or circular arrangements.[10] The obvious advantage of PATs is that they are dynamic and can be reconfigured in real time. However, they present large arrays of individual sources and therefore require elaborate support electronics to independently drive and synchronize each of those elements. That complexity and the corresponding cost of PATs scale unfavourably to larger pixel counts and higher SW. These difficulties have encouraged strategies for minimizing the number of active elements through techniques such as sparsely populated arrays and element multiplexing. Although these simplify system design, they further limit the attainable information content and the $SW_{\mathbf{H}}$.[10]

Further, the market is mainly concerned with PATs for imaging in medical and non-destructive testing. There, each element is operated in transmit and receive configuration subsequenctly. The design goals are therefore more concerned with fast switching circuits, high time resolution and high sensitivity. As a notable exception, a medical PAT system with 986 elements, which had been developed for prostate surgery, has successfully been used to reconstruct images from digital holograms.[54] A very promising result from this study was the efficacy to include a scattering object, which partially occludes the sound beam, in the computation and thus improve the focusing quality. This is a much needed capability for medical ultrasound, especially regarding HIFU and surgery, because of the highly heterogeneous landscapes in biological bodies.

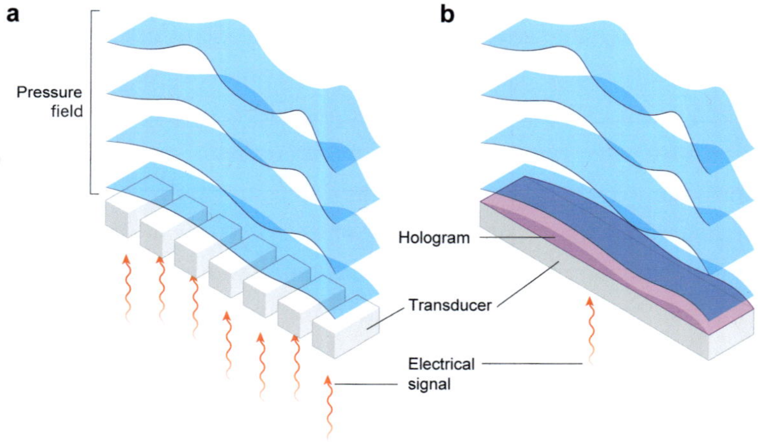

Figure 2.5: Wavefront modulation by (**a**) phased array transducer (PAT) or (**b**) acoustic phase hologram. Figure adapted from reference [55].

2.4.2 Phase Holograms

This thesis is based on another approach, similar in spirit to the kinoform in optics,[3] which ideally acts as a pure phase object with regard to the wave (Figure 2.5(b)). This is a very efficient way to modulate a wave, as in the ideal kinoform—with a smooth surface and therefore no phase steps—all diffractive orders besides the wanted first order disappear. Theoretically, diffraction efficiencies of 100 % can be achieved in this way.[56] An acoustic kinoform can be implemented as a surface profile of an extended slab of material. Due to the different speed of sound in the hologram material versus the surrounding medium a transmitted wave will be locally delayed in proportion to the material thickness. The hologram's shape is thus similar to the emitted wavefront.

The phase plate can be considered a thin element if it only modulates $\phi \in [0, 2\pi]$. In this case transmission through the element can be considered one-dimensional for each pixel position in (x, y). To further simplify the calculation, shear waves, multiple reflections and acoustic attenuation in the hologram material are disregarded. The phase lag is therefore

$$\Delta\phi(x, y) = (k_\mathrm{m} - k_\mathrm{h})\Delta T(x, y), \tag{2.6}$$

where $T(x, y) = T_0 - \Delta T(x, y)$ is the thickness of the hologram layer at position (x, y) and the subscripts of wave numbers $k = \omega/c$ refer to hologram (h) or medium (m). T_0 is a base thickness that can be freely chosen considering mechanical stability of the phase plate. In this way any wave front can easily be encoded into a hologram. Using additive fabrication (3D printers) the achievable space-bandwidth SW_H of the hologram is only limited by the printer's size or by the extent of the reconstructing beam profile. Currently available 3D-printers are suited to fabricate these holograms in the frequency range up to 7.5 MHz, which corresponds to a wavelength of 200 µm. Other fabrication methods can be employed to extend this range. It is also possible to combine holograms of different frequencies in one static element.[57]

2.4.3 Amplitude and Phase Holograms

If there is a method to modulate both amplitude and phase of a wave one can directly use the back-propagated field of a given target distribution without optimization. The capabilities of some 3D-printers to work with multiple materials on the same object enable holograms that modify both the amplitude and the phase. To this effect, a second layer made of an attenuating material is added to modulate the amplitude proportional to its thickness. The phase lag for two material layers is

$$\Delta\phi(x,y) = (k_{\mathrm{h1}} - k_{\mathrm{m}})T_{\mathrm{h1}}(x,y) + (k_{\mathrm{h2}} - k_{\mathrm{m}})T_{\mathrm{h2}}(x,y), \tag{2.7}$$

where the subscripts 1 and 2 correspond to the printed hologram materials. The amplitude change due to attenuation is

$$\ln \frac{p_0}{p_{\mathrm{in}}} = -\alpha_{\mathrm{h1}} T_{\mathrm{h1}} - \alpha_{\mathrm{h2}} T_{\mathrm{h2}}. \tag{2.8}$$

In this equation p_{in} is the incident wave from the transducer and the α are the acoustic attenuation coefficients. Solving these two equations will in most cases lead to unfeasibly thick elements, where the assumption of 1D transmission does not hold. Contrary, limiting the phase lag to $[0, 2\pi]$ limits the achievable amplitude range depending on the available 3D printing materials. Unfortunately, no sufficiently attenuating materials were available for 3D printing at the time of this writing. This concept is subject to further research.

2.4.4 Reflection Holograms

In the case of a reflection hologram the wave does not pass through the element. The local phase lag is instead caused by crevices of varying depth along the reflection surface. Reflection holograms are computed in a manner similar to transmission phase holograms. The transducer output is first propagated to the hologram plane to establish the input field across the hologram aperture. The phase lag and thickness are related as

$$\Delta\phi(x,y) = 2k_{\mathrm{m}}\Delta T(x,y). \tag{2.9}$$

Considering a perfect reflector, the phase lag only depends on the depth of the surface crevices and not on the hologram material. Reflection holograms will be used in Section 6.1 to control the field in a resonator for levitation of objects in air.

2.4.5 Optoacoustic Holograms

The photoacoustic effect can also be used to create complex waveforms suitable for holography. Two slightly different approaches to create the fields have been reported. The first approach uses an absorbing receiver whose thickness profile is shaped proportional to the wavefront.[58,59] This case is similar to the phase hologram and its $SW_{\mathbf{H}}$ is limited by the fabrication method. The second approach uses a thin, flat and highly absorbing membrane to give rise to the sound wave. There, the incident light field is shaped by a spatial light modulator (SLM) to project the desired wave amplitude distribution.[60] Each point of high light intensity corresponds to an acoustic point source. The $SW_{\mathbf{H}}$ is thus limited by the SLM's pixel count.

One advantage of optoacoustics lies in the wireless nature of the acoustic source, which is excited purely by light. However, very intense and short light pulses are required that demand complicated instrumentation. The sound pressure amplitude depends on the target temperature and heat exchange with the environment. To avoid raising the target's average temperature the laser excitation has to be pulsed. This method is not suitable for continuous waveform operation.

Chapter 3

Creating Arbitrary Acoustic Fields

This chapter presents the experimental reconstruction of ultrasound fields from 3D-printed acoustic phase holograms, starting with arbitrary 2D sound pressure images. Since a hologram technically encodes a 3D field it can project multiple images in separate planes. Finally, it is shown how sound images with amplitude and defined phase can be realized. Each section highlights the changes to the optimization algorithm that are necessary to address the particular problem. Scanned hydrophone measurements of the ultrasound fields are compared against the simulations. This chapter contains excerpts and figures from the articles "Holograms for acoustics"[1] and "Acoustic Fabrication via the Assembly and Fusion of Particles".[16]

3.1 General Workflow

The general workflow from virtual target to real image is captured in Figure 3.1. The process begins with the definition of the target image (a), where the black lines in this example define a uniform sound pressure distribution and the white background shall remain at zero amplitude. The dimensions of this image and its distance to the hologram plane will be encoded in the hologram. The transducer output and the target pressure map provide the boundary conditions for the source and image plane, respectively, as they are applied during the optimization using the

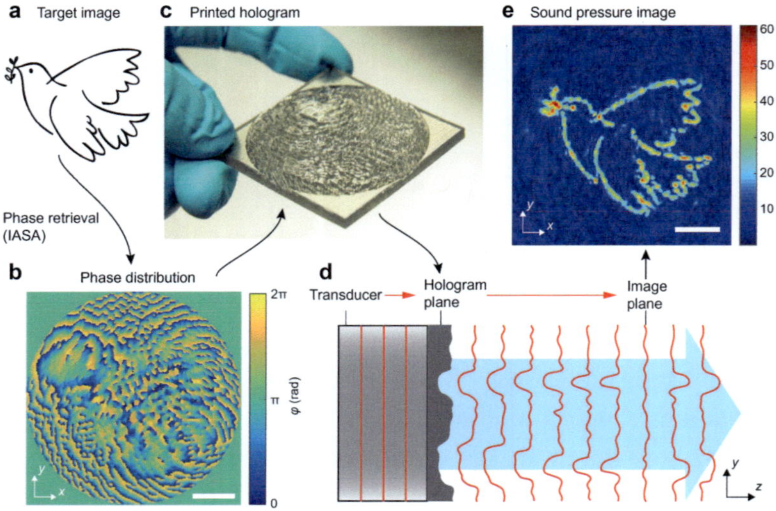

Figure 3.1: Workflow toward generating arbitrary acoustic images. (**a**) Target image amplitude used as input to the optimization algorithm. (**b**) Computed phase distribution ϕ for 2.06 MHz ultrasound. (**c**) The 3D-printed transmission hologram (50 mm side length). (**d**) The hologram converts a planar wavefront from the transducer into the desired phase distribution. (**e**) Scanned hydrophone pressure measurement in the image plane reveals the reconstructed image. All scale bars, 10 mm. Figure taken from reference [1].

IASA (Section 2.2.2). This eventually produces an optimized phase map (b) across the hologram aperture, which is then physically encoded in the thickness profile of a 3D-printed phase plate (Section 2.4.2). A photograph of the printed plate (the acoustic hologram) is shown in panel (c). To produce the real image in an experiment the printed hologram plate is put in contact with the transducer face and both are immersed in the working medium, which is de-ionized water in this work. Panel (d) visualizes how the initial plane wave emitted by the transducer is locally delayed as it passes through the hologram material. This effect results from the difference in speed of sound in the hologram material versus the surrounding medium. Ideally, the emitted wavefront equals the computed phase map. Upon propagation the wave diffracts to form the real image at the encoded distance. Scanning a hydrophone across the image plane reveals the sound pressure map, which is shown in (e). This workflow stays the same for all transmission holograms presented in this work with slight modifications to the optimization routine depending on the problem at hand.

3.2 3D-printed Phase Hologram

The optimized phase hologram can be encoded in a surface profile, as discussed in Section 2.4.2, and fabricated using a 3D printer. In this work all holograms are made by an Objet Connex 260 (Stratasys, USA) in the material VeroClear, which is a proprietary photopolymer that simulates polymethyl methacrylate (PMMA).[61] At 2 MHz the speed of sound and the sound attenuation in this material were measured to be $2424\,\mathrm{m\,s^{-1}}$ and $5.5\,\mathrm{dB\,cm^{-1}}$, respectively. The measurement is based on analysing the time of flight and amplitude of ultrasonic pulses through solid blocks of differing side lengths (USN 60, Baker Hughes, USA).

Figure 3.2 shows a photograph of a printed hologram in VeroClear material, which can be used for image reconstruction without any further processing. Because the hologram is fabricated by 3D printing, its *SW* is limited only by the printer's build size and resolution. The operating frequency used in this work of about 2 MHz, which is common in non-destructive testing and medical diagnostic ultrasound, results in a wavelength of $\lambda \approx 750\,\mu\mathrm{m}$. Counting with a modest resolution of $\lambda/2 \approx 375\,\mu\mathrm{m}$ the phase plate with 50 mm diameter contains 15 000 acoustic pixels. The fine print resolution (20 µm) in the z-direction permits to encode 100 phase steps over the full range of 2π radians of the passing wave.

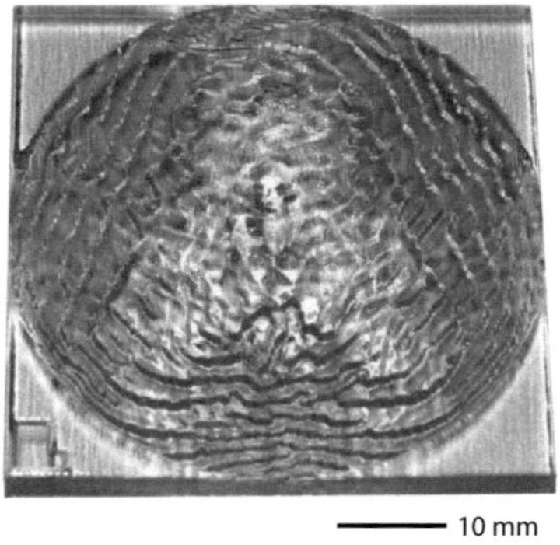

Figure 3.2: Photograph of a printed phase hologram encoding the image of the letter "A" used in chapter 5. Figure taken from reference [16].

To assess the quality of the 3D-printed hologram, X-ray computed tomography (CT) was performed using a TomoScope HV 500 (Werth GmbH, Germany) with a voxel size of 81 μm × 81 μm × 81 μm. Results are shown in Figure 3.3, where panels (a-c) present the designed thickness map before printing, the measured thickness map from the CT analysis and the difference between the two, respectively. Figure 3.3(d-f) show profile plots along different horizontal slices (constant y) through the plotted data for both the designed and measured samples. The printed hologram is in excellent agreement with the design, although it can be seen that the printer acts as a low-pass filter, smoothing out sharp edges and fine features. The X-ray CT also provides density measurements of a sample and showed homogeneous density throughout the hologram except for two small regions, which are marked by red circles in Figure 3.3(b). These defects are probably due to incomplete curing of the print material in the crevices near the object edges.

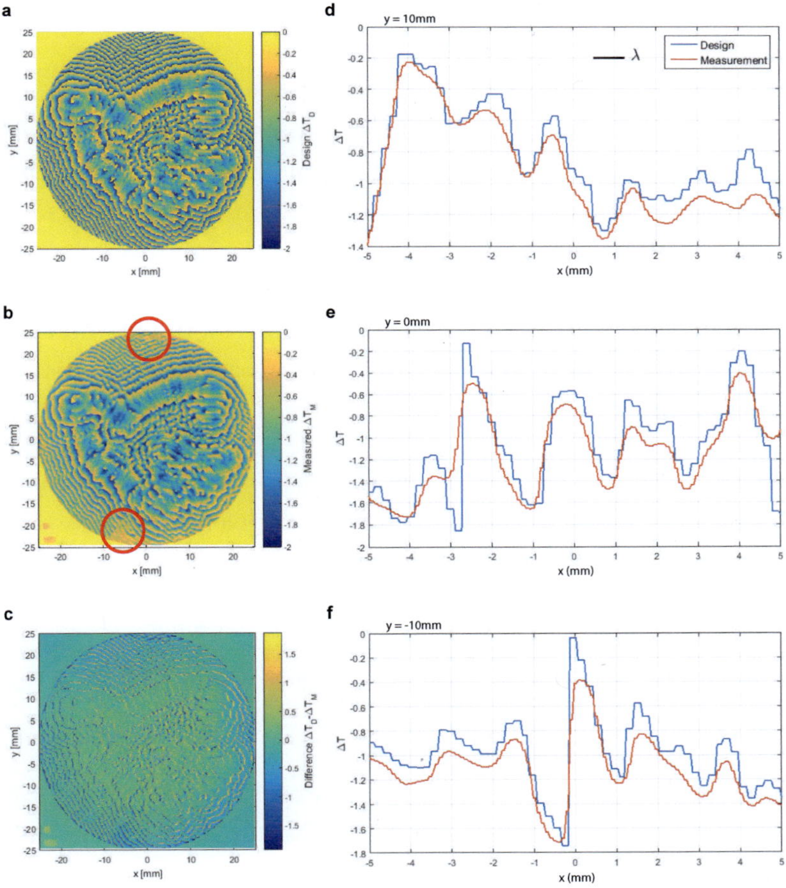

Figure 3.3: Analysis of 3D print quality. (**a**) Designed thickness map ΔT_{D} from hologram calculation, as sent to the 3D printer. (**b**) Measured thickness map ΔT_{M} of the printed hologram (X-ray CT). Red circles mark sections of inhomogeneous material density. (**c**) Difference between design (**a**) and measured (**b**) thicknesses. (**d-f**) Thickness profiles of ΔT_{D} (blue) and ΔT_{M} (red) along (**d**) $y = 10\,\mathrm{mm}$, (**e**) $y = 0\,\mathrm{mm}$ and (**f**) $y = -10\,\mathrm{mm}$. The scale bar in (**d**) shows the wavelength in water at 2.06 MHz. Figure taken from reference [1].

3.3 Experimental Setup

All experiments with waterborn ultrasound in this work are performed at room temperature (about 21 °C) in the same open-topped tank with dimensions of 600 mm × 300 mm × 300 mm (length × width × height), which is filled with de-ionized water. Correspondingly, the speed of sound for the following simulations and experiments is taken as $1484\,\mathrm{m\,s^{-1}}$. The bottom and short side walls are lined by commercially available absorber sheets (Aptflex F48, Precision Acoustics Ltd, UK) that minimize reflection of ultrasound to reduce the effect of standing waves inside the bulk medium.

Depending on the experiment the transducer and hologram can be mounted in different orientations on a manual positioning stage, which is fixed to the frame of a scanning gantry (SF600, GAMPT, Germany). The gantry allows automatic positioning of different tools in X, Y and Z-direction with micrometer precision. In a typical experiment this setup is used to scan the hydrophone through a region of interest (ROI) in front of the transducer (see Section 1.4).

There are different transducers employed throughout this work—a variety of custom built and commercial products. The custom transducer, which is used for the microparticle assembly experiments in Chapter 5, consists of a piezoceramic (PZT) plate with 1 mm thickness and 50 mm diameter, which is bonded by epoxy to a machined brass plate. The plate thickness is chosen to equal $\lambda/2$ for optimal power transfer at the projected resonance frequency of 2 MHz. The plate is mounted on a water tight case so that the brass face is exposed to water and the piezo's backside is free to vibrate in air. This design results in a transducer with narrow bandwidth.

The resonance frequency of the fabricated transducers will deviate from the designed value due to production tolerances. To accomodate this the transducers were first characterized with an impedance analyzer and then manually sweeped in the water tank in vertical, upright position. The effect of radiation pressure (Section 5.2) is clearly visible as a protruding disc on the water surface and thus a manual sweep can be used to find the optimal driving frequency for maximum power output. In this manner all fabricated transducers were found to operate in the range 1.96–2.06 MHz. The commercial transducers (e.g. A395S, Olympus, Japan) are characterized by a more uniform amplitude distribution and flat phase as expected.

A function generator (AFG 1062, Tektronix, USA) provides the driving signal, which is routed to a power amplifier (A1230, Dr. Hubert GmbH, Germany), which in turn drives the transducer. To map the acoustic fields with the hydrophone its output is connected to a lock-in amplifier (UHFLI, Zurich Instruments, Switzerland) that uses the function generator as a reference input. This combination allows accurate measurements of the complex pressure amplitude at the operating frequency including the precise relative phase between sampling points. The needle-type hydrophones have a tip size of 0.5 mm (sensitivity 545 mV MPa^{-1} at 2 MHz, Precision Acoustics Ltd., UK). The setup is controlled by custom scripts and libraries implemented using MATLAB (The MathWorks, USA).

3.4 Single Images

The most basic case of holographic reconstruction is a single image, e.g. the image of the dove from the general workflow.[1] Figure 3.4(a) shows the computed image after 20 iterations of the IASA. Its phase map was encoded and printed in an acoustic hologram as described earlier and reconstructed with ultrasound at 2.06 MHz in water. The corresponding hydrophone scan with step size $\Delta x = \Delta y = 0.25$ mm is shown in Figure 3.4(b). The reconstruction efficiencies are 59.5 % for the computed and 34.6 % for the measured pressure image.

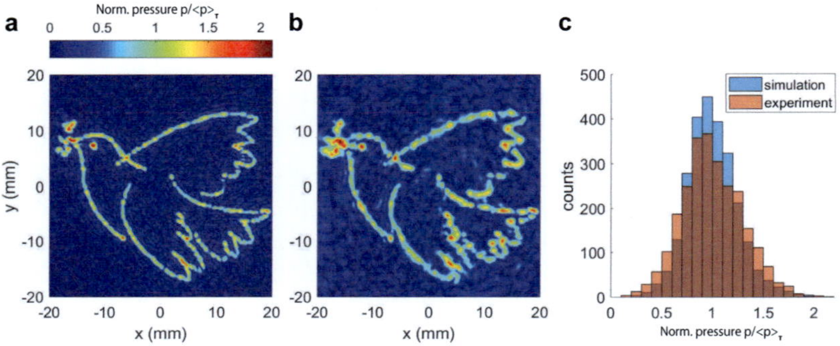

Figure 3.4: Holographic reconstruction of a single image. Pressure amplitudes are normalized by their mean value in the target region $\langle p \rangle_\mathrm{T}$. The line width of the target image is 0.8 mm (1.11λ). (**a**) Simulation. (**b**) Hydrophone scan. (**c**) Histograms showing the amplitude distributions for both, simulation and experiment. Figure adapted from reference [1].

It is evident how the simulation predicts most of the hotspots and non-uniformities in the image. Those irregularities are mainly caused by the finite extent of the hologram and a phase-only modulation. A histogram plot provides quick qualitative insight into the homogeneity of the distribution and is presented in Figure 3.4(c). Both fields are normalized by their mean amplitude values in the target region $\langle p \rangle_\mathrm{T}$. As expected, the measured field has slightly greater amplitude variance compared to the simulation.

The second example for single image reconstruction is the bicycle from reference [16], which is displayed in Figure 3.5. The overall efficiencies are 67.2 % and 44.9 % for the simulated and the measured field, respectively.

The line width for the bicycle (1.33λ) was chosen slightly thicker than for the dove (1.11λ), which probably explains the better reconstruction efficiency. In this example the amplitude homogeneiety looks slightly better in the experiment, which is probably due to the absence of some predicted high-intensity spots.

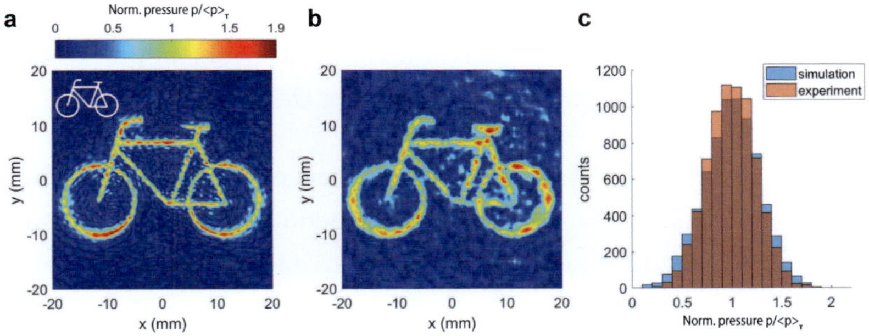

Figure 3.5: Holographic reconstruction of a single image. Pressure amplitudes are normalized by their mean value in the target region $\langle p \rangle_{\mathrm{T}}$. The line width of the target image is 1 mm (1.33λ). (**a**) Simulation. (**b**) Hydrophone scan. (**c**) Histograms showing the amplitude distributions for both, simulation and experiment. Figure adapted from reference [16].

For both examples the initial target distributions used for the hologram optimization were binary amplitude images. This results in infinitely steep edges that cannot be reproduced by a physical wave with a finite frequency bandwidth. In the reconstruction the line edges appear softened because the wave acts as a spatial low pass filter. Also the line widths for the images were chosen just slightly wider than one wavelength. More relevant than the absolute value of reconstruction efficiency is thus how well the experimentally measured pressure distributions reproduce the simulated pressure fields. In this regard both examples demonstrate good agreement between theory and practice.

3.5 Multiple Concurrent Images

An ideal hologram stores the full wave information. According to the linear wave equation this means that the evolution of the field in the volume behind the hologram is fully determined. In other words, this makes it possible to compute holograms for arbitrary 3D acoustic fields. Of course, there are physical limits imposed on the solutions. The fields have to satisfy the wave equation and obey the conservation of energy, since one cannot arbitrarily create sources or sinks in a medium.

However, a step in this direction is the concurrent projection of multiple images by one single hologram. Reconstruction happens by the same reference wave and the images appear each in separate planes along the propagation direction. This comes at the expense of quality (or homogeniety), because the limited *SW* of the hologram has to encode additional information for each additional image, but the exceptionally high *SW* of printed holograms makes it possible to maintain some reconstruction fidelity as will be shown in this section.

The extension of IASA to multiple planes is straightforward and has been inspired by a similarly modified GS algorithm.[62] There, the forward and backward projections are computed separately for each image plane. This results in individual *sub*-holograms, one for each image, which are added together after each iteration. Due to complex summation, the dominant emitter pixels for each image prevail. In the following iteration the summed hologram is propagated forward to each image plane and the cycle continues.

The following example encodes three images, the outlines of digits "1", "2", and "3", at distances 20 mm, 30 mm, and 40 mm from the hologram, respectively.[1] Figure 3.6 displays simulation (top row) and hydrophone

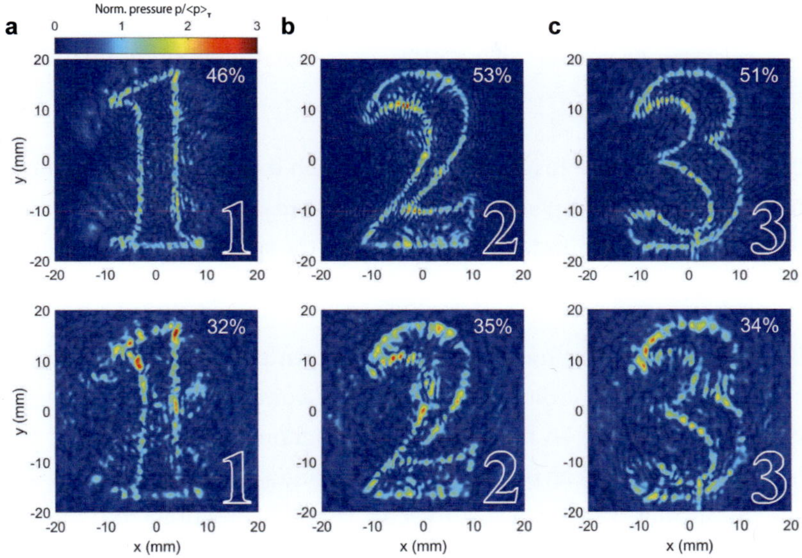

Figure 3.6: Holographic reconstruction of multiple images in separate image planes. Simulation (top row) and hydrophone scans (bottom row) for images at (**a**) 20 mm, (**b**) 30 mm and (**c**) 40 mm distance. The reconstruction efficiency for each image is shown in the top right corner. Figure adapted from reference [1].

scans (bottom row), where each plane scan took about 40 min to scan. All images were reconstructed at the same time by the same reference wave at 2.06 MHz. The individual image quality is much lower compared to a single image, but that is not directly evident from the reconstruction efficiency alone. There are visible artifacts in the images that clearly resemble shapes found in the other images, e.g. a faint appearance of number "2" in the first image plane (also called ghost image). This is not surprising because all wave packets that form part of an image have to necessarily pass the other image planes as well. Further, a wave requires some propagation distance to allow diffraction to focus to a point or, contrary, diverge towards a low amplitude background. The presence of artifacts largely depends on the image composition and distance between image planes as well as the aperture size of the hologram in relation to the images.

Figure 3.7 shows the histogram plots for each image. The experimental results show much larger variance in sound pressure amplitude. The tail towards higher amplitude values represents the speckle that are apparent in the displayed scans in Fig. 3.6.

In summary, the reproduction of 3D acoustic fields is possible but challenging. The additional information content required from the volumetric field has to be stored in a hologram of sufficient *SW*. It is beneficial if the ROI can be accessed from a larger portion of the enclosing sphere. Ideally, waves can be projected from all directions into the ROI. A first step in this direction can be a hologram with large ratio of aperture size to image distance. Ultimately, large curved holographic surfaces or multiple flat holograms positioned around the ROI are needed to create high-quality 3D fields.

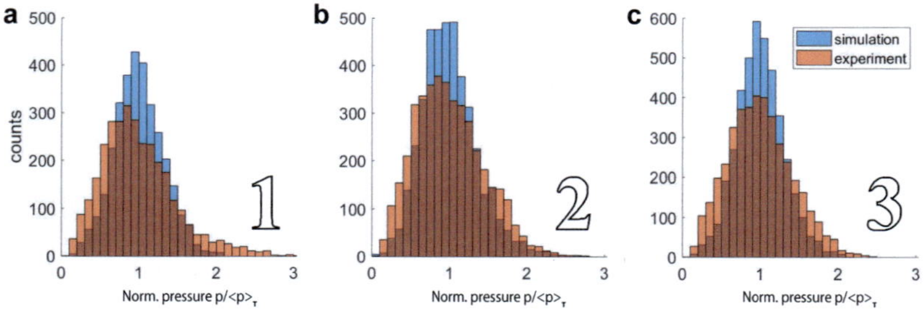

Figure 3.7: Histograms for each image plane (**a-c**) shown in Figure 3.6.

3.6 Images with Amplitude and Phase

In the previous sections the holograms had been optimized for amplitude images, each time leaving the complex phase as a free variable. Because the phase is directly linked to the propagation distance of a wave it determines which hologram pixels contribute constructively to which image pixels. In other words, a free phase allows wave contributions to arrive from different sections of the hologram and thus increases the solution space. The phase, however, is also an indicator of the wave front's shape. The relative change of phase between neighbouring points in a plane is the in-plane component of the local wave vector. Control over the phase distribution in the image thus means control over the effective wave direction at each point. As an example, a tilted plane wave would result in a linear phase gradient across the observation plane. Note, that a phase can only be defined for a non-zero pressure amplitude.

Vortex beams are good examples for wave fields with interesting phase distributions. There, the phase advances in a helical fashion around a point of singularity along the propogation axis. Well-known examples are Bessel or Gauss-Laguerre beams of order l, where the amplitude in the focus is ring-shaped and along the ring its complex phase behaves as $\phi = l\theta$. In cylindrical coordinates the angle θ is the azimuthal angle, when the z-axis is aligned with the wave propagation. The common term for l is the *topological charge* (of a vortex).[63] Section 6.2 contains a more detailed discussion of phase gradients and their application for object manipulation.

To include the more stringent boundary conditions of a fixed phase in the image, the IASA has to be modified, by weighting the hologram pixels that contribute to the proper phase relation in the image. The weights are obtained by a single back-propagation of the target image to the hologram plane and are of the form of a sub-hologram. This is in a certain way similar to the approach encoding multiple images. In each iteration these complex weights are added to the hologram as additional source constraint. The complex addition ensures that the relevant pixels in the hologram plane stay in phase, and the other pixels are further optimized.

Using this method some exemplary holograms have been obtained that encode focused vortex rings with 10 mm radius and varying topological charges advancing in counter-clockwise direction. Figure 3.8 shows experimental hydrophone scans of the reconstructed images. The phase gradients correspond to topological charges 5, 10, and 20, respectively, and the scans are in excellent agreement with the target values. As can be seen in 3.8(c), approaching larger phase gradients leads to periodic break up of the amplitude distribution, slowly approaching the appearance of a standing wave along the ring.

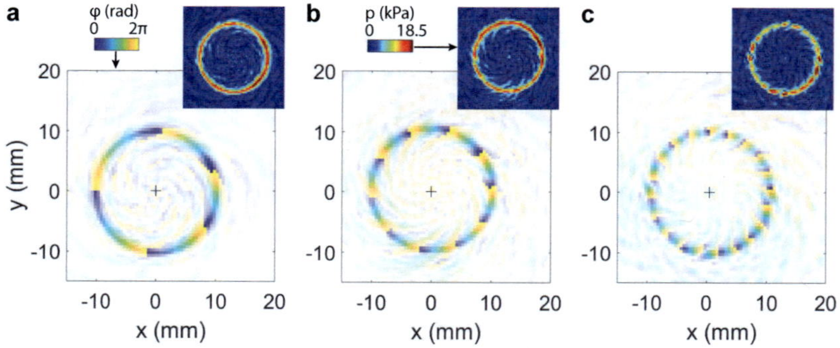

Figure 3.8: Rings (r = 10 mm) with imposed phase gradients advancing in counter-clockwise direction. Hydrophone scans for (**a**) $l = 5$, (**b**) $l = 10$, and (**c**) $l = 20$. The amplitude maps p are shown as inset and the phase plots ϕ are shaded by amplitude to enhance visibility of the tracks.

Chapter 4

Fast 3D Scanning of Acoustic Fields

In this chapter, an alternative measurement concept is presented that detects sound intensity via the heating effect it has on an absorbing membrane. The membrane is moved through the region of interest (ROI) while a thermal camera continuously records its surface temperature. This technique results in much faster scanning times for volume acquisitions. Section 4.2 has been reported in "Holograms for acoustics".[1] The remaining chapter contains excerpts and figures from the article "Fast spatial scanning of 3D ultrasound fields via thermography".[12]

4.1 State of the Art

To acquire a spatial distribution, a hydrophone has to be scanned through a set of points, a process that does not scale well to large areas or volumes. Typical acquisition rates are on the order of 1 point per second, resulting in multiple hours or even days for a full scan.[64] Long scanning times potentiate errors due to systemic changes in the environment or transducer, e.g. caused by temperature drift or evaporation of liquid. The conventional solution to reduce measurement time is to limit the scan to few relevant planes. This is sufficient for simple fields that only provide a single focus or that are symmetric about the axis of propagation.[65]

The term *acoustic holography* commonly refers to a measurement technique in the audible frequency range in air, where arrays of microphones are positioned in a way so that the field on a surface around the source can be acquired.[66] All other field variables and second order quantities, as well as surface vibrations can then be computed using holographic techniques. The much shorter wavelengths of water-borne ultrasound complicate the actual measurement setup so that the operator is left with the scanning hydrophone. Even then, using holography, one precision measurement of the complex sound pressure in a transverse plane to the wave propagation direction is sufficient to virtually compute the whole 3D field.[36] This, however, can produce significant artifacts if the acquired field contains measurement errors.

Alternative approaches rely on the *acousto-optical* effect, which refers to the change of the optical refractive index as a function of the density (pressure) of the traversed medium.[67] The effect is for instance employed in two-dimensional Fabry-Perot sensors, where the acoustic pressure changes the optical density between two parallel, reflecting surfaces of a thin film, which in turn can be detected by a scanning laser.[64,68]

A second acousto-optical technique is *schlieren imaging*, where a collimated beam of light passes through the ultrasound field and diffracts due to the varying index of refraction. The camera is placed behind a knife edge or patch stop, which blocks the undiffracted light. Thus only higher order diffracted light will reach the sensor, resulting in a 2D projection of the density variation across the measurement volume. By rotating the transducer, one can record projections at different angles and eventually reconstruct the volumetric data using the inverse Radon transform.[69] Similar results can be obtained from laser Doppler velocimetry, where a (scanned) laser beam interrogates a target after passing through the ROI. Contrary to conventional laser velocimetry measurements, the target does not move but instead the variations of the refractive index in the ROI change the effective optical path length, which can in turn be used to infer the properties of the acoustic field.[70] Difficulties arise for high amplitude sound fields when the optical phase difference in the field of view exceeds 2π.[71] Acousto-optic methods require optical access, long working distances, and specialized instrumentation.

4.2 Sound Visualization via Liquid Crystals

A simple and fast method to map acoustic fields uses the thermochromic effect of certain liquid crystals (LCs).[72] A temperature change induces a phase transition in those LC, which corresponds to a visible color alteration. Coating a sheet containing thermochromic LC with a thin sound absorbing layer causes it to display a spatial color distribution related to the locally absorbed sound intensity. Thermochromic sheets are commercially available for specific onset temperatures and temperature ranges. For example, the sheet used in the following experiment operates in the temperature range from 25–30 °C (R25C5B, LCR Hallcrest, USA). Upon heating it changes from its initial black color to a red color, then turns green, to finally appear blue above the indicated temperature range. This concept has been used to map 3D sound fields in conjunction with a camera.[1,73]

Thermochromic field visualization is demonstrated using the multi-image hologram from Section 3.5. The hologram's phase map is shown in Figure 4.1 (a) and panels (b-d) show photographs of the thermochromic sheet positioned at each image plane. As discussed above, the LC sheet has been coated with an absorbing layer (Aptflex F36, Precision Acoustics, UK) and mounted via a custom frame to the gantry of the experimental setup. In this way it can be automatically translated to each image plane

(Panels (b-d)), where the stage rests for about 3 s before a photo is recorded (EOS 60D, Canon, Japan). Comparison to the hydrophone scans in (Figure 3.6 on page 47) clearly shows the correct reproduction of high intensity regions and even reveals the ghost image of number "2" in the first plane, Fig. 4.1(b).

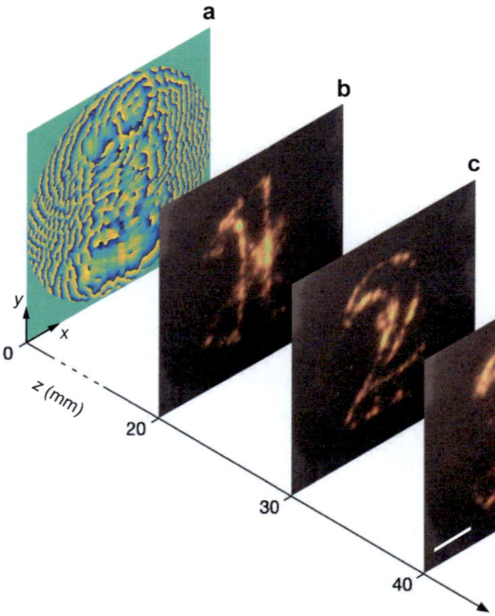

Figure 4.1: Validation of a multi-image hologram with thermochromic LC. (**a**) Phase map of the hologram. (**b-d**) The thermochromic sheet immediately reveals the local sound intensity in each image plane. Scale bar, 10 mm. See Figure 3.6 for complimentary hydrophone scans. Figure taken from reference [1].

Thermochromic visualization is a direct and fast method that does not require any special tools and is therefore very useful for quick validation and setup of holographic fields. Regions of high intensity can thus quickly be found. Orientation of the setup is not critical as long as there is optical access for lighting and imaging. However, quantitative measurement of sound intensity requires color calibration with the camera and light source and is not trivial.[73]

4.3 Thermography

4.3.1 Setup

A more quantitative solution involves recording the temperature evolution of a sound absorbing material in the acoustic field. Modern thermal cameras based on microbolometer pixels can detect infrared radiation in the wavelength range 7.5–13 μm and achieve a sensitivity of 50 mK. The inherent drawback of all indirect thermal measurements is, of course, their preference of large temperature differences and thus low sensitivity towards sound intensity.

A schematic of the experimental setup and the data analysis is displayed in Figure 4.2. The ultrasound transducer is positioned in a water tank and pointing up. The measurement device is an open cylindrical cup with a thin membrane forming the bottom surface, which is made of a sound absorbing material. A thermal camera is mounted at a fixed distance from the membrane. An automated stage translates camera and membrane through the ROI, e.g. along the z-coordinate. The cup ensures that the top surface of the membrane is always exposed to air and allows direct viewing by the camera. When the sound field is turned on it propagates through the liquid, reaches the membrane, and then partially couples into the membrane where it is absorbed. This causes a local temperature rise at the surface of the membrane, which is detected by the camera. Each camera image shows the temperature distribution in an entire plane (XY). By moving the cup and membrane with a known speed while recording a sequence of thermal images a 3D volume distribution of the recorded temperature maps can be constructed, which in turn indicates the absorbed sound intensity. Calibration of the membrane and camera with a known ultrasound field allows estimation of the sound pressure in the same volume.

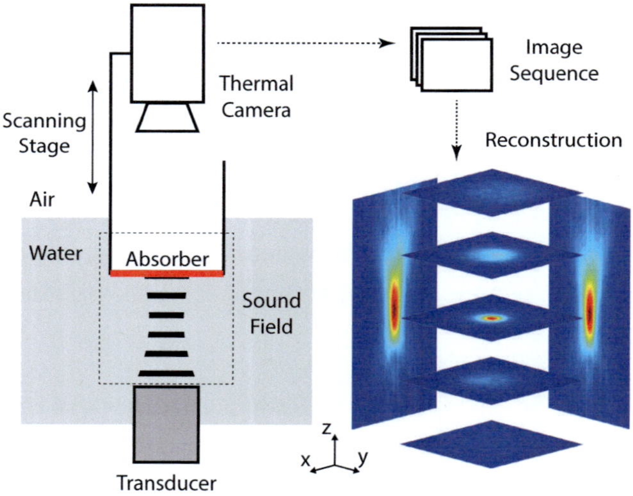

Figure 4.2: Schematic of the thermographic setup. A membrane is locally heated by the ultrasound field and a thermal camera continuously records its surface temperature. Continuous scanning along z produces a sequence of thermal images, which correspond to the volumetric sound intensity map. Shown are exemplary XY-planes and projections of the central XZ- and YZ-plane for a single focal spot. Figure taken from reference [12].

The membranes used in the experiments were fabricated from two different rubber materials. The first material is an isobutylene isoprene rubber (IIR) (Reichelt Chemietechnik GmbH + Co., Germany) with a thickness of 1.0 mm. The second material is optimized for ultrasound absorption in the lower Megahertz range and available as a two-component fluid (Aptflex F36, Precision Acoustics Ltd., UK), which contains scattering particles embedded in a soft polyurethane matrix. Thin films of Aptflex F36 are cast on a 0.1 mm thick polyethylene substrate to provide mechanical stability. The total thickness of both layers was then measured to be 0.43 ± 0.02 mm using a digital microscope (VHX-6000, Keyence, Japan). The distance between the thermal camera (A65, FLIR Systems, USA) and the membrane in this setup is fixed to 200 mm.

4.3.2 Process Analysis

The performance of this method depends on several parameters including the thermal and mechanical properties of the absorbing membrane and its thickness, the scanning speed and the thermal camera. A simplified geometry is displayed in Figure 4.3(a) for a material with acoustic attenuation coefficient α_s, thermal conductivity κ_s, and specific heat capacity $C_{p,s}$. The top surface is exposed to air while the bottom surface is in contact with the water bath at room temperature. Coupling of an acoustic wave into a thin material layer depends on its mechanical properties in contrast to the liquid medium and its thickness in relation to the wave-length. Analyzing this problem leads to the reflection coefficient[12]

$$r = \frac{p_-}{p_+} = \frac{Z_s - Z_m - (Z_s + Z_m)\exp(2H(i\omega/c_s - \alpha_s))}{Z_s + Z_m - (Z_s - Z_m)\exp(2H(i\omega/c_s - \alpha_s))}, \tag{4.1}$$

where p_+, p_- are the incident and reflected pressure waves at the interface from liquid medium towards the membrane layer, respectively, H is the thickness, and ω is the angular frequency of the sound wave. $Z_x = \rho_x c_x$ is the acoustic impedance with density ρ_x, speed of sound c_x, and the subscripts refer to the (s) attenuating layer and (m) liquid medium. From this follows the coupling coefficient

$$\gamma = 1 - \left| r^2 \right|, \tag{4.2}$$

which relates the transmitted (and therefore lost) sound intensity to the incident sound intensity. It is plotted in Figure 4.3(b) for approximations to the materials used in the following experiments. As expected, a thicker membrane absorbs more total power with distinct peaks corresponding to multiples of half the wavelength. The remaining sound intensity, which is reflected, will travel further through the setup and may eventually return to the membrane. A potential problem are multiple reflections between transducer and membrane. These reflections will introduce systematic errors that depend on the length of the open space above the transducer. The ideal membrane exhibits a coupling factor of $\gamma = 1$ to absorb the entire wave. The coupling coefficients for the two membranes used in the experiments at 2.25 MHz are estimated at $\gamma = 0.4$ for the Aptflex membrane (marked **A** in Fig. 4.3(b)) and $\gamma = 0.3$ for the IIR rubber (**B**).

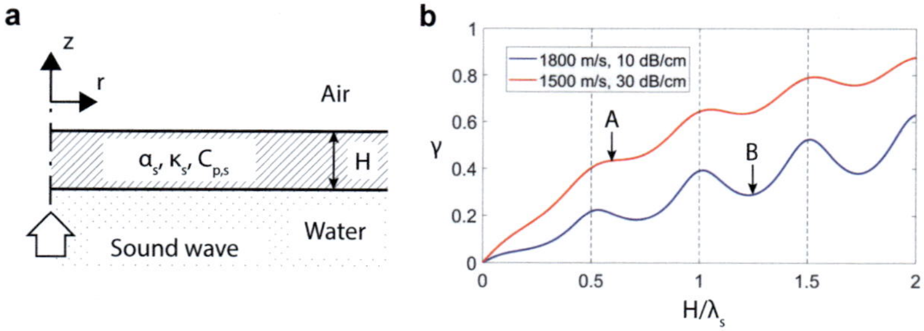

Figure 4.3: (**a**) Model of the system. (**b**) Coupling factor γ for the two different materials. Figure adapted from reference [12].

The wave portion that couples into the membrane layer is absorbed and converted to heat, which can be expressed as[74]

$$q = -\nabla I_{\mathrm{s}} = 2\frac{\alpha_{\mathrm{s}}}{\rho_{\mathrm{s}} c_{\mathrm{s}}} \langle p_{\mathrm{s}}^2 \rangle, \tag{4.3}$$

where the brackets $\langle \cdot \rangle$ denote a time average.

When the transducer is turned on, the surface temperature in the absorbed spot quickly rises and then slowly converges to a steady state with the time constant τ, which can be approximated as[75]

$$\tau = \rho_{\mathrm{s}} C_{\mathrm{p,s}} H^2 / \kappa_{\mathrm{s}}. \tag{4.4}$$

This time constant should be minimized to improve response time and therefore spatial accuracy in a continuous scan. The optimal conditions of low τ (fast response) and high γ (high attenuation) suggest that thin membranes made from highly absorbing materials are ideal for thermographic recording of ultrasound fields.

A corresponding model has been developed in Matlab using the Partial Differential Equation Toolbox (The Mathworks, USA). This allows transient and steady state solutions to be computed for this problem.

4.3.3 Calibration

The setup can be calibrated by reference to a known sound field that has been properly characterized using a precision method, i.e. a hydrophone scan. A convenient reference field is a single focus. A fresnel lens with focal length 30 mm has been made for a large area immersion transducer (A395S, Olympus Scientific Solutions, USA) operating at 2.25 MHz.

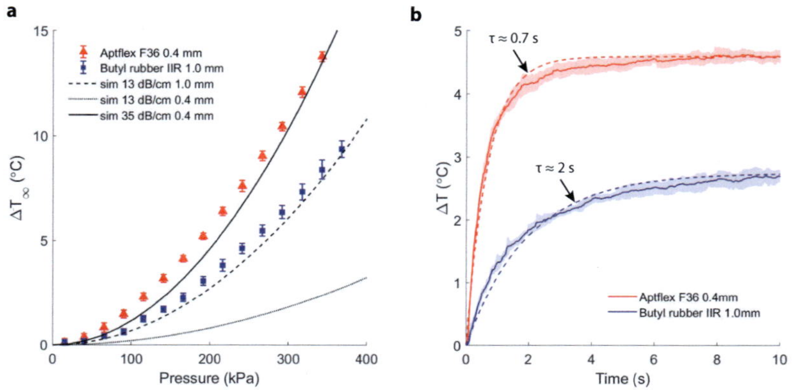

Figure 4.4: (**a**) Steady state peak temperature plotted against peak pressure in the focus. (**b**) Average step response curves from 5 recordings per material (solid lines — mean, shaded regions — one standard deviation). Dashed lines show fits of exponential functions to the data and their respective time constants are annotated. Figure taken from reference [12].

Two measurements are necessary to complete the calibration dataset. The first establishes the relation between driving voltage of the transducer and sound pressure amplitude in the focal spot. The peak sound pressure is acquired using a calibrated needle hydrophone (0.5 mm, sensitivity 545 mV MPa^{-1}, Precision Acoustics, UK). In a second experiment the membrane is positioned in the focal plane and the thermal camera records the steady state temperature in the same focus versus the driving voltage. In both cases the points of maximum pressure and temperature were found by a combination of automated scanning and manual inspection.

The two measurement results are combined and displayed for each material in the chart of Figure 4.4(a). Simulation results for three different general rubber materials are added for comparison. The parameters are chosen to correspond to the two rubber materials that were used in the experiments. Those parameters are $\kappa_s = 0.2\,\mathrm{W\,m^{-1}\,K^{-1}}$, $C_{p,s} = 1900\,\mathrm{W\,s\,m^{-1}\,K^{-1}}$, $c_s = 1800\,\mathrm{m\,s^{-1}}$, $\rho_s = 1100\,\mathrm{kg\,m^{-3}}$ with varying thickness and attenuation coefficients as labeled. The data clearly shows the expected quadratic dependence between temperature and sound pressure from Equation 4.3. Those curves serve as calibration data in the following experiments to recover sound pressure amplitude from scanned temperature profiles.

4.3.4 Measurement Performance

The dynamic performance of the system can be inferred from its step response. For this measurement the transducer is turned on at a specific point in time and the temporal evolution of the surface temperature is recorded. The step responses for multiple sequential runs are shown in Figure 4.4(b) and the thinner material shows faster settling time towards the steady state. The plot also shows fitted exponential functions and their respective time constants for each case. A smaller time constant is beneficial for an instrument because it follows more readily dynamic changes in the field. The measurement noise can be expressed as a noise equivalent pressure (NEP).[64] To quantify the NEP, the temperature of single thermal camera pixels has been recorded over a 5 s period without ultrasound exposure. Three times the standard deviation over this period was found to be approximately 0.1 °C for all recordings, which corresponds to an NEP of 23.6 kPa for Aptflex and 34.1 kPa for the IIR rubber using the previously obtained calibration set. For comparison, the broadband noise (100 MHz) of the hydrophone system is rated at 60 µV by the manufacturer leading to an NEP of 330 Pa. In conjunction with a lock-in amplifier at 2.25 MHz (input range set to 200 mV, i.e. 367 kPa) an NEP of just 1.6 Pa was measured.

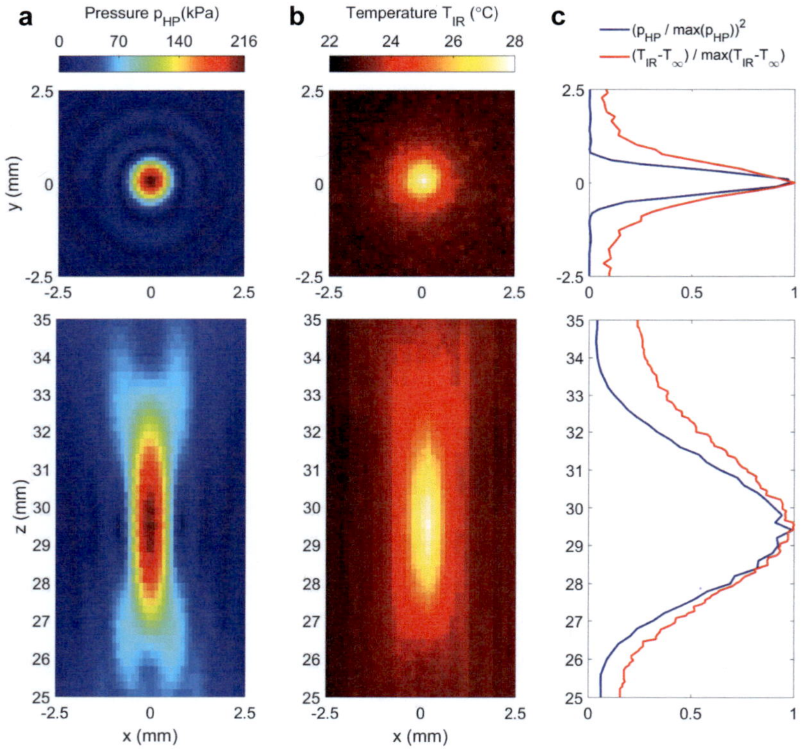

Figure 4.5: Scans of an ultrasound focus at 2.25 MHz and transducer driving voltage $U_{\mathrm{TD}} = 15\,\mathrm{V(p\text{-}p)}$. Displayed are slices through the focus, XY (top) and XZ (bottom). (**a**) Pressure amplitude from hydrophone scans. (**b**) Recorded temperature distribution from a thermographic scan of the same field using a membrane made of 0.4 mm Aptflex F36 rubber ($\tau \approx 0.7\,\mathrm{s}$) Scanning velocity $v_z = 1.8\,\mathrm{mm\,s^{-1}}$. (**c**) Squared pressure and temperature profiles (along $x = 0$) normalized by their respective maximum values in the measurement. Room temperature, $T_\infty = 22.1\,°\mathrm{C}$. Figure taken from reference [12].

Using a calibration target the size of each pixel in the thermal images was found to correspond to a length of 0.13 mm. The resolution in the scanning direction, however, is v_z/f_{cam}, where v_z is the scanning speed and f_{cam} the frame rate of the camera. Typical values in this setup are $v_{\mathrm{z}} = 1.8\,\mathrm{mm\,s^{-1}}$, $f_{\mathrm{cam}} = 30\,\mathrm{s^{-1}}$, resulting in a z-resolution of 0.06 mm per pixel. The uncertainty following such a discretization is $\pm 1/2$ of the step size. Microbolometer detectors such as employed in this thermal camera exhibit an impulse response to the IR signal, which the manufacturer reports as 12 ms in the datasheet. This is much smaller than the time constants τ for heating the membrane and accordingly the latter defines the dynamic response of the absorber in the z-direction,

$$g(z) = \frac{1}{\tau}\exp\left(\frac{-z}{\tau v_{\mathrm{z}}}\right). \tag{4.5}$$

Figure 4.5 shows slices of measured temperature distributions for the better performing material (Aptflex F36) and the reference sound field measured by a hydrophone. The thin and highly absorbing membrane gives a clear picture of the overall dimensions and location of the focal spot. It is noticeable how the focus in the thermographic scan is extended along the positive y-axis. Following the dynamic analysis, the systematic error in the positive scanning direction can be expressed as $\tau v_z = 0.7\,\mathrm{s} \times 1.8\,\mathrm{mm\,s^{-1}} = 1.26\,\mathrm{mm}$.

4.3.5 Results

The calibration data can now be used to rapidly scan an unkown field. For this experiment a phase plate will be used that projects two foci (F1 and F2) in different positions and distances from the aperture. Its phase map is shown in Figure 4.6(a).

As described earlier the thermal camera obtains a set of thermal images during a continuous scan and the calibration data from Figure 4.4(a) serves to convert the temperature change to a sound pressure amplitude. The resulting pressure map is displayed in Figure 4.6(b) along with the hydrophone scans around each region of interest displayed in Panels (b1) and (b2). The hydrophone measurements reveal peak pressures of 127 kPa (F1) and 122 kPa (F2). The peak pressure amplitudes acquired by the thermographic method with the Aptflex membrane are 115 kPa (F1) and 130 kPa (F2), where the variations are comparable to the previously calculated NEP of 23.6 kPa.

The thermographic scan covered a total volume of 30 mm×30 mm×60 mm in about 33 s providing a rapid assessment of the 3D sound pressure field compared to conventional methods. Figure 4.6(c) illustrates volumetric information of the foci via pressure isosurfaces around each point.

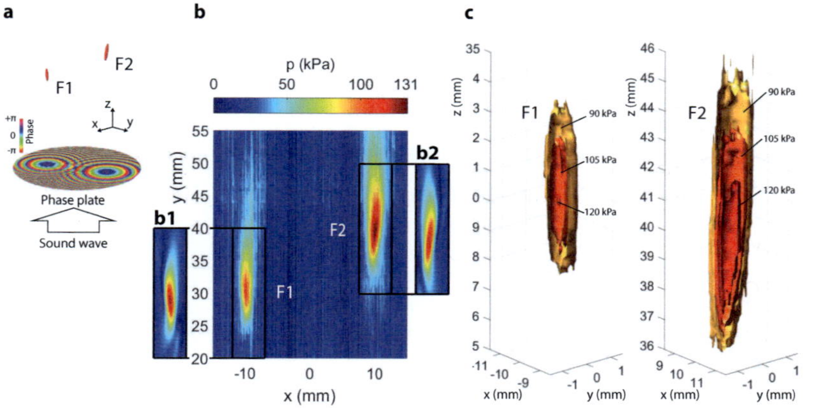

Figure 4.6: (**a**) Phase plate for generating two foci, F1 and F2. (**b**) Central slice ($y = 0$) of reconstructed pressure distribution from a thermographic scan. The sub-panels (**b1, b2**) show hydrophone scans of the respective regions. (**c**) 3D pressure isocontours around F1 and F2 obtained from the thermographic data of (**b**). Figure adapted from reference [12].

4.4 Conclusion

In summary, this chapter introduced thermography as a methodology to rapidly scan 3D volumes of ultrasound fields within seconds and with a spatial resolution in the sub-millimeter range. The method relies on mechanically translating a sound absorbing sheet through a body of water. While moving, the membrane is continuously exposed to the ultrasound field and its changing surface temperature is captured by a thermal camera. The resulting image stack is subsequently arranged to a volume temperature map and calibration data serves to convert the temperature to sound pressure amplitude.

The sensitivity and precision of the method rely on and can be controlled by the characteristics of the membrane material and its thickness, the scanning speed, and the resolution of the thermal imaging system. Using a thin ultrasound absorber sheet made from Aptflex F36 and polyethylene, the temperature map of a 3 cm $\times$ 3 cm area with 10^4 pixels at a lateral resolution of 0.13 mm can be acquired at a time. Scanning the membrane in the third dimension then maps a volume of 54 cm^3 in just 33 s.

Thermography is not as sensitive or accurate as a hydrophone measurement. Its strength, however, lies in instant visualization of 2D ultrasound fields and thus fast scanning of volumetric fields. There it is orders of magnitude faster than hydrophone scanning. It is much simpler to implement than acousto-optical setups and does not require tomographic reconstruction. It is thus a convenient tool to complement emerging field synthesis techniques (e.g. acoustic holograms), but also dynamic applications of ultrasound.

Chapter 5

Acoustic Assembly and Fabrication

The acoustic hologram presents a major advance in shaping ultrasound fields and the following sections demonstrate how these structured fields can push microparticles into arbitrary arrangements. Once the particles are assembled in the desired shape they can be fixed, for example by chemical reactions. A new fabrication method based on acoustic assembly is presented. This chapter contains excerpts and figures from the articles "Holograms for acoustics"[1] and "Acoustic Fabrication via the Assembly and Fusion of Particles".[16]

5.1 State of the Art

The forces that result from sound waves incident on interfaces have already interested Lord Rayleigh and since then fascinated several generations of scientists.[13] Over the last century the field has progressed immensely and sound waves have been applied to manipulate particles from microscopic to millimeter sizes. Its contact-free nature, wide scalability of wavelength and intensity over orders of magnitude, paired with low attenuation in most solids and liquids makes it a viable choice to explore future manipulation and fabrication, especially with view on biological matter.[76]

The majority of previous work in acoustic particle manipulation is based on acoustic resonators, which arrange many particles along extended traps that coincide with either nodes or antinodes of the standing pressure waves.[14] This concept is well established in microfluidics where resonances with high quality factor can be achieved due to precision microfabrication techniques.[27] The straightforward implementation is a 1D standing wave across a flow channel. Particles or cells moving with the flow can thus be pushed perpendicular to the preeminent drag of the moving liquid. Careful design combined with a flow split downstream allows separation of particles and even sorting by size or compressibility.[77]

Attempts to move this concept towards higher complexity of the patterns change the resonator's aspect ratio from elongated channels to flat two-dimensional cavities, with one or multiple transducer elements placed along the side walls.[78,79] Due to multiple reflections inside the chamber these methods are limited to highly symmetric fields corresponding to resonant modes. The concept has also been extended to a 3-dimensional rectangular cavity excited from 5 sides by individual transducers. This created a dynamically reconfigurable metamaterial, where the transmission of certain frequency bands could be controlled by adjusting the spacing between trapped particles, which in turn was set by the trapping frequencies.[80] However, due to the geometry of the setup the achievable structures are limited to cubic lattices. One solution to overcome the limitations imposed by the cavity's geometry has been demonstrated in the form of a flat cavity lined by 64 elements that were each highly damped by attenuating backing. The transducer elements were then controlled as a phased-array and able to trap and move individual particles within the acoustic cell.[81]

The forces on particles resulting from travelling waves are comparatively less important, since the largest force component, which is in the direction of wave propagation, is usually not wanted. An intensity gradient is required to raise the lateral force components, demanding a strongly focused beam. In this way, trapping of polystyrene (PS) beads (with diameter 3–100 μm)[82] and lipid droplets (126 μm)[83] has been achieved using transducers with high-numerical aperture (NA) driven at very high frequencies in the range 30 to 400 MHz. Here, the particles and droplets were situated on a membrane that prevented forward motion with the acoustic wave. A breakthrough in 2016 was the 3D trapping of single PS spheres in the center of a focused vortex beam generated by a high-power PAT.[84] More recently a similar setup for acoustic tweezers has been presented using a different phase singularity effect, which the authors called a twin trap.[85] Ultrasonic manipulation has been investigated as an improvement to other fabrication methods, for example to control the make up of composites in 3D printing. The acoustic radiation force (ARF) was used to position and align fibers in a resin while it is cured via stereolithography.[86,87] However, with all the existing methods it has not been feasible to generate acoustic fields for particle assembly in truly arbitrary shapes.

5.2 The Acoustic Radiation Force (ARF)

Interaction of a sound wave with an obstacle changes the wave's momentum, which leads to an effective *Langevin radiation pressure* acting on the obstacle's surface.[88] Modes of interaction include absorption of the wave but also reflection, refraction, and scattering. The analysis in the latter cases is more complicated than perfect absorption, but necessary to understand particle trapping and manipulation by sound.

When a wave with amplitude $|p|$ hits a perfectly absorbing target the radiation pressure[89]

$$p_{\mathrm{rad}} = E_{\mathrm{ac}} = \frac{1}{2}\frac{|p|^2}{\rho_{\mathrm{m}} c_{\mathrm{m}}^2} = \frac{I_{\mathrm{c}}}{c_{\mathrm{m}}}, \tag{5.1}$$

where ρ_{m}, c_{m} are the density and speed of sound in the surrounding medium, respectively, E_{ac} is the acoustic energy density and I_{c} the characteristic intensity from Eq. (1.14). In the case of perfect reflection it is twice that value, meaning $p_{\mathrm{rad}} = 2E_{\mathrm{ac}}$. The effect of the ARF

$$F_{\mathrm{rad}} = p_{\mathrm{rad}} A = \frac{P}{c_{\mathrm{m}}} \tag{5.2}$$

on a target surface A is the underlying principle of how ultrasonic power balances measure the acoustic power of sound sources. For example, a fully intercepted and absorbed sound beam in water ($c_{\mathrm{m}} = 1484\,\mathrm{m\,s^{-1}}$ at 21 °C) results in a force of $0.67\,\mu\mathrm{N\,mW^{-1}}$.[89]

5.2.1 Travelling Plane Waves

The ARF on particles, whose dimensions are on the order of the acoustic wavelength or smaller, mainly results from scattering of the incident sound wave. Computation of the ARF is therefore based on the solution to the scattering problem. This work follows the established route to compute the ARF via partial wave expansion, which is covered by references regarding the general problem,[35,90] the approximation towards small particle sizes and the radiation potential,[91] and the extension towards arbitrary fields.[92] Only the final results will be shown here and the reader is kindly directed to the cited references for details. The incident and scattered field are expressed in spherical harmonics where the coordinate system's origin is put at the particle center. Once the scattered field is obtained the force can be found by integrating the Brillouin radiation stress tensor over a closed surface around the scatterer.

For a plane travelling wave scattered at a spherical particle there is only one non-zero component of the force, which is aligned with the wave propagation, in this case along z. In an inviscid fluid[35]

$$F_z = -\lambda_m E_{ac} \sum_{n=0}^{\infty} (n+1)\,(A_n + A_{n+1} + 2A_n A_{n+1} + 2B_n B_{n+1}), \tag{5.3}$$

where λ_m is the wavelength in the medium. The scattering coefficients A_n, B_n depend on the size and material composition of the scatterer. Figure 5.1 shows the computed radiation force acting on spherical particles in water as a function of their radius a for two selected materials, polystyrene (PS) and polydimethylsiloxane (PDMS). The general trend in both cases is for larger particles to feel stronger forces, but particles of all sizes experience a force of the same sign in the positive z-direction. Scattering resonances are visible where the particle size approaches the wavelength.

For small particles ($a \ll \lambda_m$) Equation (5.3) can be reduced to[91]

$$F_z\Big|_{a \ll \lambda_m} \approx \frac{4}{9}\pi a^2 E_{ac}\left(f_1^2 + f_1 f_2 + \frac{3}{4} f_2^2\right)(k_m a)^4, \tag{5.4}$$

where a is the particle radius and k_m the wavenumber in the medium. This approximation is plotted in red against the exact solution in Figure 5.1. The approximation somewhat overestimates the force for larger particle sizes.

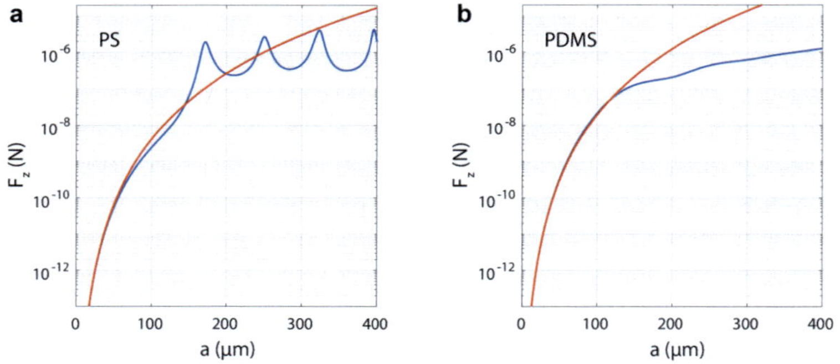

Figure 5.1: Radiation force on a spherical particle exposed to a travelling plane wave ($f = 2\,\text{MHz}$, $|p| = 100\,\text{kPa}$, $\lambda_m \approx 750\,\mu\text{m}$). Approximation for $a \ll \lambda_m$ shown in red. (**a**) Polystyrene (PS), (**b**) Polydimethylsiloxane (PDMS).

The two parameters introduced in Eq. (5.4) are the monopole coefficient[35]

$$f_1 = 1 - \frac{\rho_m c_m^2}{\rho_s c_{l,s}^2} \cdot \frac{3(1-\nu_s)}{1+\nu_s}, \tag{5.5}$$

and the dipole coefficient

$$f_2 = \frac{2(\rho_s - \rho_m)}{2\rho_s + \rho_m}, \tag{5.6}$$

which are used to define the *acoustic contrast factor*[15]

$$\Gamma_a = \frac{1}{3} f_1 + \frac{1}{2} f_2. \tag{5.7}$$

Coefficient f_1 is a function of the Poisson ratio ν_s, which relates transverse expansion to axial compression of a material. In this way it also relates the propagation speeds for compression (i.e. longitudinal) and shear (i.e. transverse) waves in the solid particle material. They are denoted $c_{\mathrm{l,s}}$ and $c_{\mathrm{t,s}}$, respectively, and[33]

$$\frac{c_{\mathrm{l,s}}}{c_{\mathrm{t,s}}} = \sqrt{\frac{2(1-\nu_{\mathrm{s}})}{1-2\nu_{\mathrm{s}}}}. \tag{5.8}$$

For most rigid solids $\nu_{\mathrm{s}} = 0.3\ldots0.4$ and thus $c_{\mathrm{l,s}}/c_{\mathrm{t,s}} = 1.9\ldots2.5$. However, soft materials including polydimethylsiloxane (PDMS)[93] are characterized by $\nu_{\mathrm{s}} \rightarrow 0.5$ and the ratio increases multifold as the speed of shear waves approaches zero.

5.2.2 The ARF in a 1D Standing Wave

The same particle located in a one-dimensional standing wave experiences a radiation force depending on its actual position relative to the pressure nodes and anti-nodes. For a distance h away from the nearest anti-node (i.e. pressure maximum) the force is[35]

$$\begin{aligned} F_z = {} & \lambda_{\mathrm{m}} E_{\mathrm{ac}} 2\sin(2k_{\mathrm{m}}h) \times \\ & \sum_{n=0}^{\infty} (n+1)(-1)^n \left(-B_n + B_{n+1} + 2A_n B_{n+1} - 2B_n A_{n+1}\right), \end{aligned} \tag{5.9}$$

where a positive force points towards the node. It can be seen how F_z reaches a maximum at $h = \lambda_{\mathrm{m}}/8$, and equally drops to zero approaching either node or antinode. The approximation for small particles ($a \ll \lambda_{\mathrm{m}}$) in this case is[91]

$$F_z\Big|_{a\ll\lambda_{\mathrm{m}}} \approx \lambda_{\mathrm{m}} E_{\mathrm{ac}} 2\sin(2k_{\mathrm{m}}h)\Gamma_{\mathrm{a}}, \tag{5.10}$$

where magnitude and direction of the ARF are determined by the acoustic contrast Γ_{a}.

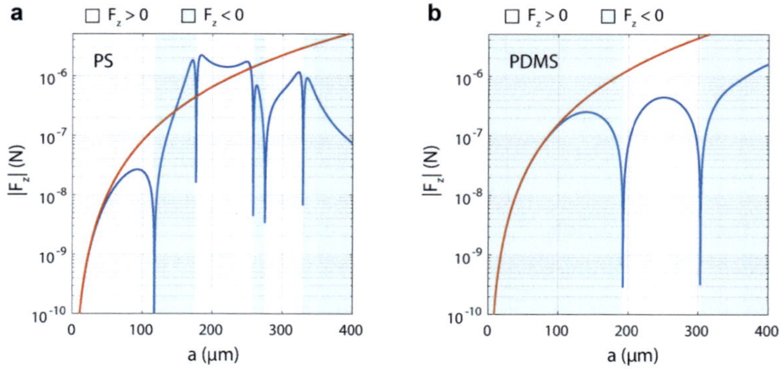

Figure 5.2: Maximum radiation force F_z on a spherical particle in a 1D standing wave ($f = 2\,\mathrm{MHz}$, $|p| = 100\,\mathrm{kPa}$, $\lambda_\mathrm{m} \approx 750\,\mu\mathrm{m}$). Positive force is towards the closest pressure node. Shaded regions indicate negative force values of the blue curve, i.e. direction inversion towards the anti-node. Approximation for $a \ll \lambda_\mathrm{m}$ shown in red. (**a**) Polystyrene (PS), $\Gamma_\mathrm{a} = 0.17$. (**b**) Polydimethylsiloxane (PDMS), $\Gamma_\mathrm{a} = -0.34$.

In Figure 5.2 both exact calculation and approximation are shown in in blue and red, respectively, for two exemplary materials. At small particles sizes F_z is positive (negative) as is expected for a particle of positive (negative) acoustic contrast. Sign inversion takes place for larger particles in discrete regions, which cannot be predicted by the approximation. It should be noted that in these regions the particle diameter $2a$ is larger than the maximum travel distance towards the nearest (anti-)node, in this example $\lambda_\mathrm{m}/4 \approx 180\,\mu\mathrm{m}$, and thus the practicability of this effect for particle manipulation is limited. However, it serves as a good comparison to the travelling wave case.

5.2.3 The Radiation Potential

Lev Gor'kov noted that for small particles ($a \ll \lambda_\mathrm{m}$) exposed to acoustic fields with sufficiently large spatial variation (i.e. excluding travelling plane waves) it is convenient to define the ARF as the gradient of a radiation potential U,[91]

$$\mathbf{F} = -\nabla U, \tag{5.11}$$

where

$$U = \pi a^3 \left(\frac{1}{3} f_1 \frac{|p|^2}{\rho_\mathrm{m} c_\mathrm{m}^2} - \frac{1}{2} f_2 \rho_\mathrm{m} |u|^2 \right). \tag{5.12}$$

The appropriate term for $\mathbf{F}$ in this case is *gradient force* as opposed to scattering force since it is independent of the propagation direction of the incident wave. The formulation of a radiation potential is especially useful for calculating the force map in resonator geometries, where often the whole field can be expressed in analytic form. Otherwise it requires the computation or measurement of the sound pressure in all relevant dimensions. The potential vanishes if $|p|^2$ and $|u|^2$ are uniform in space, in which case other terms dominate and lead back to Eq. (5.3).

5.2.4 The ARF for Arbitrary Fields

The ARF on large particles in arbitrary sound fields, such as those resulting from holographic reconstruction, requires a little more effort. Sapozhnikov et al. derived an elegant way to compute the radiation force directly from the AS of the acoustic wave.[92] As described in Section 1.2 the AS decomposes the field into elementary plane waves propagating in different directions throughout the positive half-space. The scattered field for an elementary plane wave is known from Eq. (5.3). The total scattered field resulting from an arbitrary incident field is then the sum of all elementary solutions contained in the AS. The radiation force can then again be found by integration of the radiation stress over a closed surface around the scatterer. Since the partial solutions provide the force on a scatterer located at the origin, the AS needs to be shifted accordingly across all sampling points to obtain an extended force map. This approach is used in the following sections to compute force maps from measured hydrophone data sets.

5.3 Parallel Particle Assembly

From section 5.2.3 it is clear that for fields with sufficiently large spatial variation, the force is dominated by the gradient of the acoustic field variables. In this case the acoustic contrast factor Γ_a determines the force magnitude and direction. Considering a reconstructed sound field a particle with $\Gamma_a > 0$ will be pushed against the sound pressure gradient away from the focal lines and points. A negative contrast is thus advantageous to assemble particles with a positive projected image. However, reaching $\Gamma_a < 0$ in water requires more compressible and less dense matter. Most solid materials do not satisfy these requirements. One exception is PDMS, which is consequently used as the model particle to explore acoustic assembly and fabrication.[94]

To demonstrate particle manipulation via ARF the setup from Section 3.3 has been expanded by a custom particle chamber, which can be mounted at a fixed distance from the transducer and hologram. The chamber is made from a 60 mm diameter acrylic tube, which is cut to 20 mm length and covered at its top and bottom by 100 μm thin sheets cut from polyethylene terephthalate. Those windows are acoustically transparent, which means they allow the acoustic wave to pass through the container with minimal reflection (estimated at less than 10 %) and therein maintain its travelling-wave character. The schematic of the setup is shown in Fig. 5.3(a), where the transducer is immersed in the water tank and oriented pointing up. The cell is positioned so that its top window coincides with the plane where the reconstructed image forms. When the ultrasound transducer is turned on, the particles are pushed upwards (away from the hologram) in the direction of the acoustic wave until they hit the top window. There the z-movement is restrained and the particles migrate within the plane and along the pressure gradient to rearrange in the shape of the positive acoustic image. When the sound is turned off the particles disperse, falling under the influence of gravity. The acoustic assembly is therefore only a temporary state. Formation of solid permanent objects therefore requires an additional step, which is discussed below.

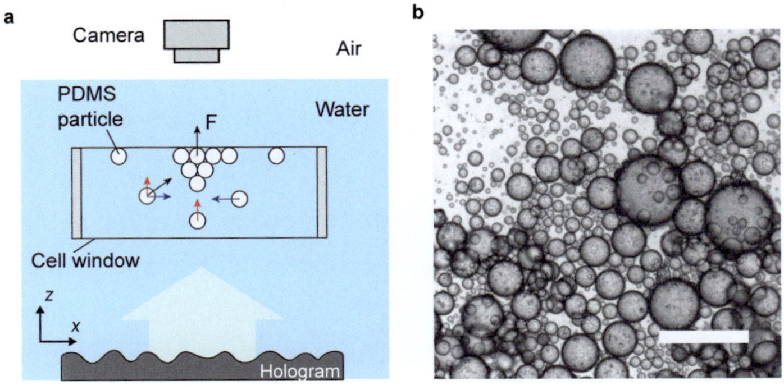

Figure 5.3: (**a**) Schematic of the experimental setup and assembly mechanism: the particles are pushed up into the transparent barrier by F_z and gathered in-plane by the lateral acoustic gradient forces F_x, F_y. (**b**) Microscopy image of the PDMS particle suspension used in trapping experiments. Scale bar 300 µm. Figure adapted from reference [1].

5.3.1 Preparation of Soft Silicone Particles

The PDMS microparticles were fabricated in an emulsion using a technique adopted from the work of Johnson et al.[95] The silicone precursor is commercially available (Sylgard 184, Dow Corning, USA) in two parts, base and curing agent. They were mixed at the standard ratio of 10 : 1 and added to a water bath with 1 wt% non-ionic surfactant (Pluronic F108, Sigma Aldrich, USA). The emulsion was heated to 60 °C, continuously agitated at 3600 min^{-1} using a high-speed emulsifier (Turaxx T 18, IKA-Werke, Germany) and then left to cure overnight. A microscope image of the suspension can be seen in Figure 5.3(b). Due to the fabrication process the PDMS particles are polydisperse with sizes up to 400 μm diameter. PDMS has a density of 1045 $kg\,m^{-3}$, bulk modulus of 1.1 GPa and the speed of sound is 1020 $m\,s^{-1}$.[94] In water these properties lead to an acoustic contrast factor of $\Gamma_a = -0.34$. Consequently PDMS particles are attracted to pressure maxima.

5.3.2 Results

The sound image chosen to demonstrate particle trapping is the dove from Section 3.4. Figure 5.4(a) shows a version of the same hydrophone scan with scaled amplitude to match the conditions in the experiment. The manipulation experiments are performed using a custom ultrasound transducer. It consists of a piezoceramic disc bonded to a thin brass plate and driven at its first thickness resonance frequency, which was measured to be 2.06 MHz. The electrical impedance at resonance is relatively low with 2.9 Ω. Particle trapping requires to drive the transducer with a higher voltage amplitude $U_{tx} = 12.5\,V$ and thus electrical power $P_{tx,el} = 27\,W$, which results in considerable heating of the device. To minimize errors from thermal drift and other effects over the long duration of a hydrophone scan, the measurement is conducted with the transducer driven at low power with $U_{tx} = 5\,V$ and thus $P_{tx,el} = 4.3\,W$. A manual check ensured linear scaling between applied voltage amplitude and the transducer output. Finally, the measured pressure map was scaled by a factor of 2.5 to match the higher power conditions used in the trapping experiment. This pressure map is used as input for the following ARF computations.

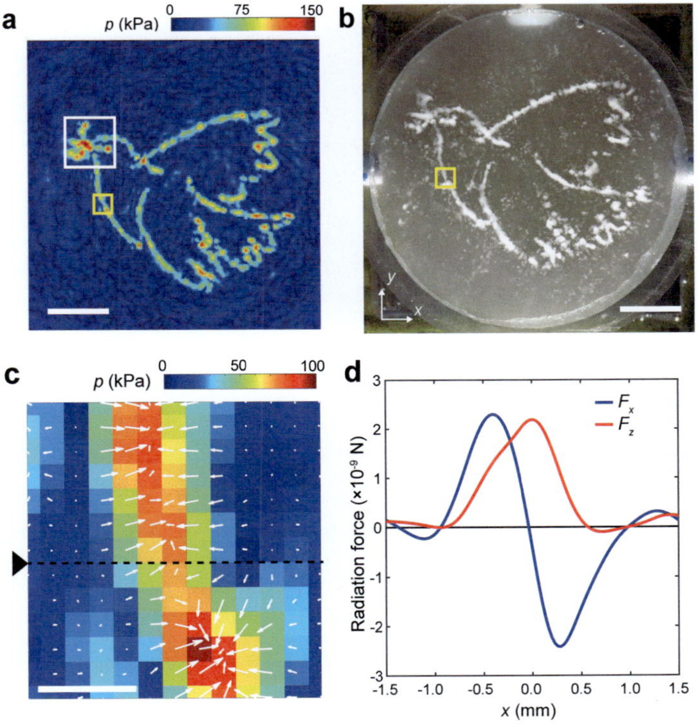

Figure 5.4: 2D particle assembly. (**a**) Hydrophone scan of the acoustic pressure field (amplitude scaled by factor 2.5). (**b**) Photograph of the particle assembly during ultrasound exposure. Scale bars 10 mm. The yellow square marks the position of the detailed view in (**c**), which further shows computed radiation force vectors for 150 µm diameter PDMS particles. Scale bar, 1 mm. The dashed line indicates the position of the profile plots in (**d**). F_x, F_z are calculated from the angular spectrum decomposition. Figure adapted from reference [1].

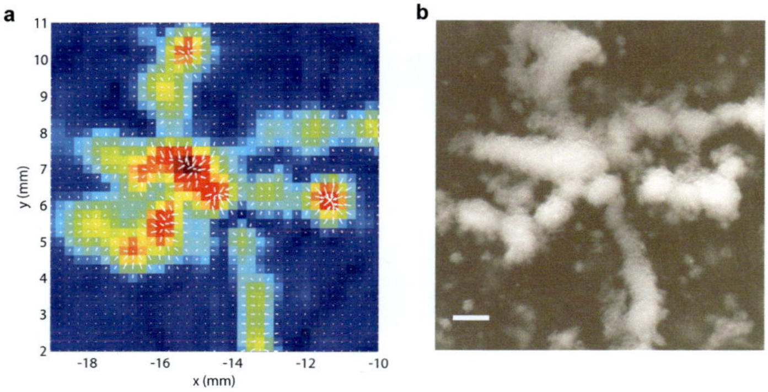

Figure 5.5: Detailed section of Fig. 5.4 (white box) showing (**a**) force vector map and (**b**) the corresponding photograph with trapped particles. Scale bar 1 mm. Figure taken from reference [1].

When the sound is turned on the particles move as described earlier and after about 5 s they display the final particle image as seen from the top in Figure 5.4(b). Individual particles can still be seen moving within the high intensity regions. An exemplary force map for the section marked by a yellow box is shown in Figure 5.4(c) along with a profile plot in panel (d). Another detail section from the head of the dove, marked by a white square, and the corresponding pressure and force maps can be seen in Figure 5.5. It is remarkable how well the particle clusters represent the measured speckles and inhomogeneities in the pressure map. In most trapping experiments the formation of streaming vortices could be observed in the fluid cell. The fluid moves upwards in the center along with the particle motion and downwards near the sidewalls. Excess particles that are not trapped in the image are recirculated and eventually sediment out along the perimeter of the bottom window. These vortices seem to be generated by regular acoustic streaming but also fluid entrainment due to massive particle motion.

5.4 Acoustic Fabrication

Acoustic forces can trap and move particles in place and serve as a tool for contact-free assembly of whole particle colletives into arbitrary shapes. However, these formations are temporary and disband once the acoustic forces vanish. By adding a fixation step that permanently binds the particles to each other a new production technique can be imagined—Acoustic Fabrication.

5.4.1 General Scheme

Figure 5.6 shows the general scheme of acoustic assembly and fabrication. The process starts with a homogeneous dispersion of particles in a suitable phase, e.g. a liquid. Then an ultrasound field is created which directs particles to preferred positions inside the volume (Fig. 5.6(b)). The advantage of using acoustic holograms is that the sound pressure distribution can be controlled with high fidelity and independent of the container geometry. A global trigger, such as light, heat, or microwave radiation can be used to fix the particles to each other. In the following demonstration the particles are fixed via a polymerization reaction initiated by ultraviolet (UV) light. To achieve this the particles had to be functionalized with an initiator and dispersed in a solution of precurser polymer. Upon illumination with UV light, the particles are bound together while the host polymer solution remains unaffected. The final product (Fig. 5.6(c)) is a fixed particle composite that can be taken out of the suspension. The main advantage of this fabrication method is that it fixes assemblies as a whole, including potentially entire 3D constructs.

Figure 5.6: General scheme of acoustic fabrication. (**a**) A homogeneous dispersion is exposed to a structured ultrasound field (**b**), which defines trapping regions for particles. (**c**) UV light triggers the fixing reaction and thus permanently fixes the particles to create the object. Figure taken from reference [16].

5.4.2 Functionalization of Silicone Particles

After assembly the particles need to be permanently joined together. However, PDMS has no chemically reactive groups on the surface, so that chemical conjugation or functionalization is difficult.[96] This prevents direct crosslinking of PDMS particles. One solution to this problem is the mechanical deposition of chemical initiator onto the particle surface. This can be accomplished by solvent-induced swelling.[97]

Figure 5.7 illustrates the functionalization scheme of initiator-laden PDMS particles. First, the PDMS particles were fabricated as described in Section 5.3.1 and subsequently dried. The particles were then immersed in a solution of acetone and 1 wt% photoinitiator (2,2-dimethoxy-2-phenylacetophenone (DMPA), > 98 %, Santa Cruz Biotechnology, USA), which shows strong absorption of UV light from 310–390 nm.[98] The particles were left under stirring for 30 min. Acetone leads to moderate swelling of the beads and therefore facilitates the uptake of the initiator into the PDMS matrix.[99] Afterwards the swollen particles were again dried at room temperature, which physically leaves the initiator molecules on or close to the particle surface. The prepared particles were resuspended and kept in the aqueous solution of 1 %(wt) Pluronic F108 to prevent aggregation.

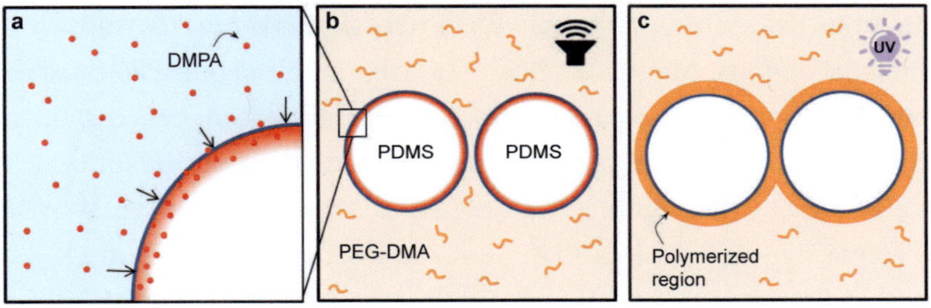

Figure 5.7: Illustration of the preparation of PDMS-DMPA particles to be UV-polymerized after acoustic assembly. (**a**) PDMS particles are swollen with DMPA photoinitiator in an organic solvent. (**b**) The PDMS-DMPA particles are distributed in aqueous solution containing PEG-DMA and assembled by the acoustic forces. (**c**) Under UV-irradiation, the polymerized volume of PEG-DMA grows outward from the particle surface and eventually binds the assembly. Figure taken from reference [16].

5.4.3 Fabrication of Rigid Objects

Immediately before the experiment, the particle suspension was first dried and then resuspended at 5 vol% in an aqueous solution of 15 vol% PEG-DMA (M_n 750 g mol^{-1}, Sigma-Aldrich, USA). A higher concentration of PEG-DMA on the particles leads to better mechanical strength, but it also enhances the diffusion of DMPA into the solution, such that the two effects need to be balanced. The suspension was then filled into the particle cell and positioned above the transducer in the water tank. Similar to the assembly process, the distance between hologram and chamber is set so that the top membrane coincides with the image plane of the hologram. Once the transducer is turned on and the ultrasound field established, the particles assemble within a few seconds in the shape of the image. Some streaming could be observed, which seems to be partially induced by both acoustic streaming and by the collective particle motion in the surrounding liquid. This results in vortices in the chamber that entrain and recirculate particles that are not trapped in the focal regions. Moderate streaming helps with the particle transport and does not negatively affect the acoustic fabrication process. Over time these recirculated particles deposit along the bottom sidewall of the chamber where there is less acoustic intensity. When the assembly is stable, typically in less than 20 s, the UV light (Omni Cure LX 400, Lumen Dynamics, Canada) is manually switched on for up to 40 s. This defines start and duration of the photopolymerization, where higher UV intensities accelerate the curing.

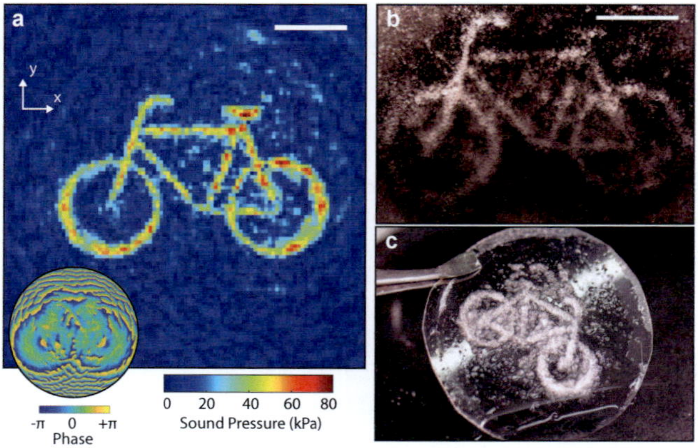

Figure 5.8: Fixation of silicone particles. (**a**) Measured pressure field in the image plane ($z = 30\,\text{mm}$). Corresponding hologram (50 mm diameter) is shown at bottom left of the image. (**b**) Particle assembly after 20 s of ultrasound exposure. (**c**) Fixed structure after one single exposure (20 s) and a postcuring step (60 s). All scale bars 10 mm. Figure taken from reference [16].

The assembled structures on the surface after one exposure cycle were found to be only a few layers of particles thick. Figure 5.8(b) shows such a structure, which is thin and not fully continuous. After removing the top membrane from the chamber the object is subjected to a postcuring and rinsing step, which also releases the object from the membrane. The bicycle shown in Figure 5.8(c) was not mechanically stable without the membrane support.

To arrive at mechanically stable products, more manageable images were chosen depicting letters of the alphabet with line widths of up to 5 mm. To further increase the mechanical strength, the assembly and curing process was repeated up to three times using the same solution. Between the steps, the chamber was removed from the tank and gently shaken by hand to redistribute the particles. In each assembly step additional layers were able to form around the existing fixed particles and partially filling gaps of the uneven structure. Figure 5.9 shows different letters of the alphabet that were fabricated using this method.

Figure 5.9: Photograph of mechanically stable and self-supporting objects in the form of letters A, B and C. Scale bar 10 mm. Figure taken from reference [16].

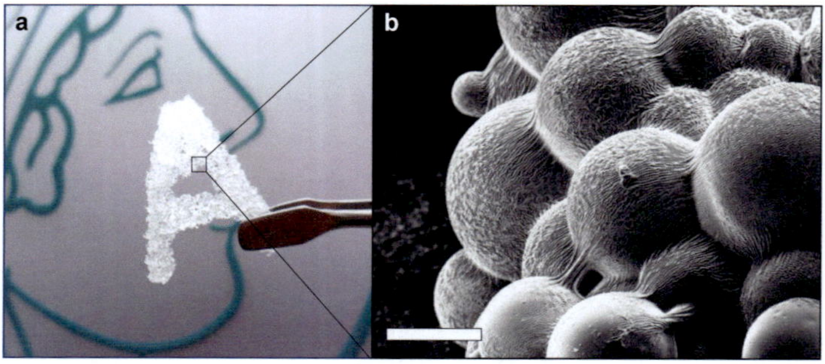

Figure 5.10: (**a**) The fabricated letter "A" held by tweezers. (**b**) SEM image of PDMS particles homogeneously covered and connected by the PEG-DMA coating. Scale bar 100 μm. Figure taken from reference [16].

For multi-layer fabrication, it was found benefitial to decrease the exposure times with every iteration from 25, 20, to 15 s at a constant UV intensity of 27.5 mW cm^{-2}. Afterward, the chamber was taken from the tank, and the top membrane removed. At this point the particle structure remained attached to the membrane, which was placed on a flat surface with the structure facing up and then post-cured for 60 s in air using the same UV lamp. The post-curing step increased the mechanical rigidity of the structure. Any unused polymer solution was subsequently rinsed off with deionized water. The structures were finally left to dry at ambient conditions.

The letter "A" is shown in Figure 5.10(a) held up by tweezers to demonstrate its self-supporting mechanical strength. A close-up view (SEM) of the crosslinked particle composite is displayed in Figure 5.10(b), which shows continuity of the PEG-DMA polymer component covering the PDMS microparticles.

Chapter 6

Acoustic Manipulation of Objects

This chapter discusses alternative ways of object manipulation through sound waves. The first part considers acoustic levitation in standing waves in air. Conventional levitation setups trap particles at nodal points along the axis between the transducer and a reflector. Replacing the latter with an acoustic hologram breaks the axial symmetry and creates laterally separated trapping points. The second part of the chapter presents a way to use a static hologram, but obtain a seemingly dynamic form of object manipulation exploiting the interplay of acoustic forces and surface tension to confine and propel objects. The acoustic hologram encodes a field of such intricacy that both the amplitude and complex phase can be defined along a path in the image plane and thereby set the object's trajectory. Due to the cyclic nature of phase this concept can move a particle indefinitely on a circular track as long as the soundfield is active. This chapter contains excerpts and figures from the article "Holograms for acoustics".[1]

6.1 Levitation in Air

6.1.1 State of the Art

Stable trapping of liquid droplets in an acoustic standing wave has been observed as early as 1933.[100] In these experiments alcohol was added to the surfaces of quartz crystals vibrating in the ultrasonic range 67–91 kHz. The alcohol atomized to a mist, which immediately migrated to the pressure nodes. The authors then observed coalescence of larger droplets from the mist that remained in the traps until evaporated, which lasted up to 2 min.

The relevant field variables for acoustic levitation are shown in Figure 6.1 for the case of a 1D standing wave, where the reflecting boundaries at top and bottom of the schematic are one wavelength apart. One of the boundaries is the transducer face driving the resonant field. Most acoustic levitation systems are of this form with variation in reflector distance and shape. For example, a curved reflector can improve lateral confinement on the axis. Displayed are the distributions of pressure p and particle velocity u, the Gor'kov radiation potential U_{rad} and the resulting acoustic radiation force F_{rad}.

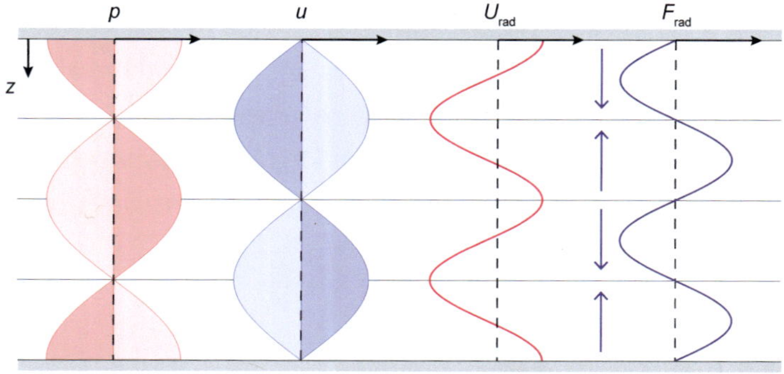

Figure 6.1: Acoustic radiation force in a 1D standing wave. Shown are sound pressure p, particle velocity u, radiation potential U_{rad} and acoustic radiation force F_{rad}, where blue arrows indicate the force direction for particles of positive acoustic contrast factor $\Gamma_{\mathrm{a}} > 0$.

Relative to a gaseous working medium *all* solid and liquid particles are rigid and dense. The monopole and dipole coefficients are approximately one and the acoustic contrast factor $\Gamma_a = 5/6$. All particles levitated in air therefore experience the radiation force pushing them towards the closest pressure node ($p = 0$). Balancing Equation (5.11) with the particle weight cancels the particle's radius and results in approximate pressure amplitudes, 2500 Pa to trap a water droplet and 250–400 Pa for a particle of expanded PS. Note, that these numbers are only rough estimates because the radiation potential is valid only for $a \ll \lambda$.

Acoustic levitation has been used for non-contact handling of small objects, where the adhesion force in conventional grippers typically starts to become problematic.[101] Another application is in container-free chemistry, which is beneficial for certain crystallization processes.[102] One study used an ultrasonic horn opposing a reflector to levitate liquid droplets in a wide size range up to 5 mm.[103] A more sophisticated arrangement of 1D levitators can move droplets laterally and thus start and observe chemical reactions without the influence of container walls.[104] Levitation and trapping of solid tungsten spheres (its density is $18.9\,\mathrm{g\,cm^{-3}}$) has also been shown and even small living animals have been trapped.[105,106] However, over the course of all studies the instrumentation has hardly changed.

Using PATs created from low-cost ultrasonic distance sensors it was recently made possible to extend the capabilities of acoustic levitation in air and dynamically change the nodal positions.[107] More recently, this approach has been improved and shown to be capable of creating more complex fields such as bottle beams or foci with phase singularities, which enabled tweezing of expanded PS by a single sided actuator.[108,109]

6.1.2 The Acoustic Reflection Hologram

One main hurdle for working with ultrasound in air is the low acoustic impedance of gases. Those result in poor coupling of waves from the transducer into the environment. The common approaches of using horns or resonators to match the transducer with the environment do not integrate well with the concept of a transmission hologram.

However, an ultrasonic wave that is already propagating through air can be manipulated via a reflection hologram (see section 2.4.4). Similar to the transmission hologram the local phase delay can be encoded in a surface profile. The depth and thus the pathlength of a round trip upon reflection can be set for each pixel.

Computation of a reflection hologram has to account for multiple reflections between the transducer and hologram surface. This decreases the effective degrees of freedom and the resulting fields are much more simple compared to the transmission holograms projecting a field into an infinite body of water.

6.1.3 Results

As discussed in Section 2.4, the complexity of a wavefront is described by the space-bandwidth product SW. To increase the information content it is therefore necessary to increase the extent of the hologram with respect to the wavelength. The commonly available ultrasonic transducers for air are designed for relatively low frequencies 20–100 kHz and without the space bandwidth SW in mind. The transducer elements used in reference [109], for example, are driven at 40 kHz and their active element is about the size of the wavelength in air, which is 8.6 mm.

The following levitation experiments employ the same electronic instrumentation as described earlier, but instead with a transducer (Multicomp MCUSD40, Farnell element14, UK) operated at 100 kHz. In air this corresponds to a wavelength of 3.4 mm. The transducer housing's diameter is 40 mm but the effective crystal diameter was measured to have a diameter of 20 mm. This restricts the space-bandwith product so that only very simple fields are possible.

In the experiment the transducer was placed opposite a reflection hologram with an air gap of 25.7 mm. Experiments were conducted with the cavity oriented vertically or horizontally. In the vertical arrangement the transducer can be either at top or bottom position. Trapping lifetimes were around 30 s, mostly limited due to thermal stability and resonance drift of the transducer, which the manufacturer does not recommend for continuous excitation.

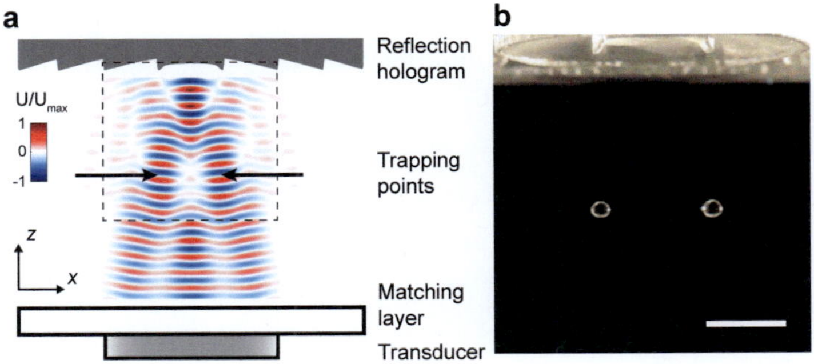

Figure 6.2: Levitation via an acoustic reflection hologram. (**a**) Radiation potential field to determine trapping sites. (**b**) Photo of two water droplets stably trapped at the indicated positions in (**a**). Scale bar, 5 mm. Gravity is acting downwards. Figure taken from reference [1].

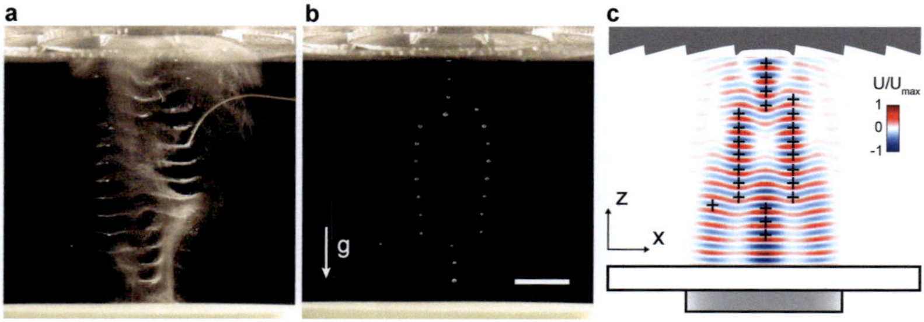

Figure 6.3: (**a**) Water mist, aerosolized at the transducer face, (**b**) coalesces into droplets trapped in the acoustic pressure nodes. (**c**) Calculated radiation potential, where the + signs indicate traps with droplets (**b**). All scale bars, 5 mm. Gravity is acting downwards. Figure adapted from reference [1].

Figure 6.2(a) shows a schematic of the experimental setup (side view) with the reflection hologram on top and the transducer positioned below. Also shown is the computed Gor'kov radiation potential inside the cavity, where the blue sections represent the minima that solid and liquid particles are drawn to. The presence of the hologram modifies the 1D geometry to have two lateral trapping sites. Droplets can be placed and stably trapped at these sites as shown in Figure 6.2(b). The water droplets shown in the photograph were manually loaded into the traps with a precision syringe. An alternative way to load liquids into the trapping sites is shown in Figure 6.3. It is the same method as reported by Bücks et al.[100] Due to the high amplitude of vibration a water film on the transducer surface will be readily aerosolized (panel (a)). The water mist is then pushed by the ARF to the pressure nodes, where it coalesces to form larger droplets (panel (b)).

Solid particles with higher density, up to that of Aluminium ($\rho = 2.7\,\mathrm{g\,cm^{-3}}$), can be levitated with the same setup as shown in the photographs of Figure 6.4. The irregular samples were cut from larger material blocks.

Lighter objects, such as expanded polystyrene (Figure 6.5(a)) or hollow glass microspheres (Figure 6.5(b,c)), can even be trapped by the lateral component of the radiation force as demonstrated in a horizontal arrangement. The collective glass spheres are compressed into very thin pancake-like forms.

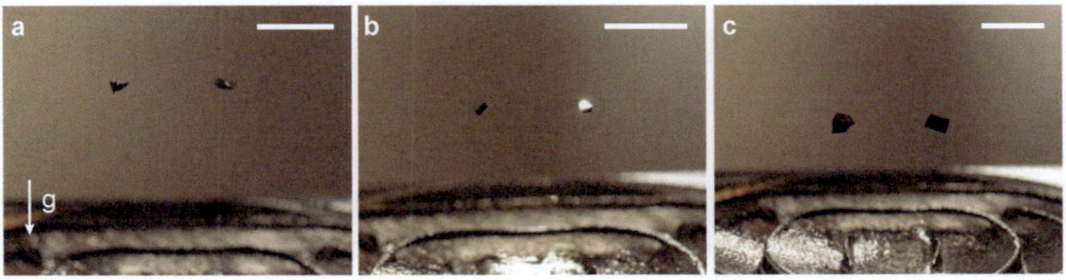

Figure 6.4: Photos of levitated objects in air. (**a**) Aluminium, (**b**) silicon and (**c**) lithium. All scale bars, 5 mm. Gravity is acting downwards in all photos. Figure taken from reference [1].

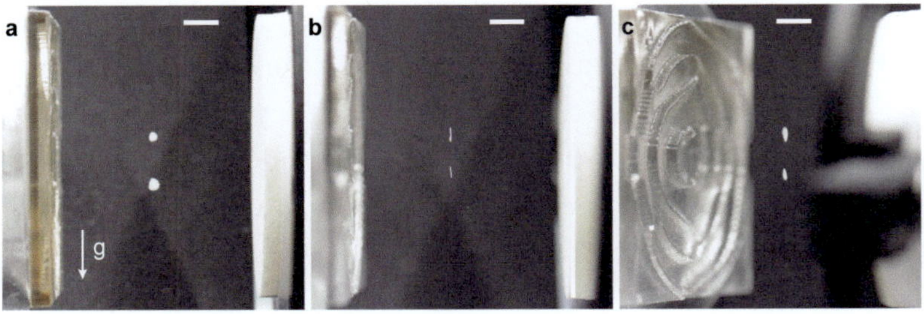

Figure 6.5: Photos of levitated objects in air in horizontal configuration. The transducer is visible to the right and the transmission to the left. (**a**) Two trapped beads of expanded polystyrene. (**b, c**) Hollow glass microspheres are trapped in thin pancake shaped nodal regions. All scale bars, 5 mm. Gravity is acting downwards in all photos. Figure taken from reference [1].

6.2 Object Propulsion by Phase Gradients

The preceding chapters and sections mostly ignored the actual wave propagation direction with regards to particle manipulation. The forward component was at most used to push particles into the region of interest and then canceled by barriers, where the assembly happened largely due to gradient forces. The complex phase, which encodes the propagation direction and the spatial development of a wave, was freely optimized during the hologram calculation. However, fixing the phase in the design process—and especially assigning a monotonically changing variable to it— enables different forms of object manipulation by the radiation force. A phase gradient $\mathbf{q}$ in an observation plane, with normal vector $\mathbf{n}$, is simply the in-plane component of the beam's wave vector $\mathbf{k}$,[110]

$$\mathbf{q} = \mathbf{k} - (\mathbf{k} \cdot \mathbf{n})\mathbf{n}. \tag{6.1}$$

Therefore it is also the corresponding component of the radiation force in that plane. Unlike the gradient force, whose sign depends on the material composition of the particle in relation to the surrounding medium (i.e. the acoustic contrast), the direction of the force resulting from a phase gradient is the same for all particles. In other words, for a plane wave incident at angle α

$$||\mathbf{q}|| = \sin\alpha \, ||\mathbf{k}||. \tag{6.2}$$

A distinct feature is the cyclical nature of the phase in $[0, 2\pi]$ for continuous excitation. This essentially allows one to define a closed, e.g. circular, phase gradient, which is impossible to achieve for an amplitude based gradient. Phase control can thus perpetually move objects in a *static* sound field.

6.2.1 State of the Art

A related mechanism is the transfer of orbital angular momentum from helical acoustic waves to levitated or floating objects,[63,111] or to microparticles trapped in highly focused beams.[112] Those vortex beams are characterized by a circular phase relation with azimuthal angle θ that is quantized to integer multiples of 2π per revolution, as in the term $l\theta$, where $l \in \mathbb{Z}$ is called the topological charge of the beam. This circular phase results in a central region $l/k = l\,\lambda/2\pi$, where the field is not defined and the sound pressure equals zero. Vortex beams are a class of non-diffractive beams since this singularity persists over long distances, surrounded by rings of high sound pressure amplitude.

As shown in Section 3.6 the acoustic hologram can project vortex fields of high topological charge (up to $l = 20$). The high space-bandwidth of a printed hologram allows control of the complex phase of the wave field beyond vortex beams. Advanced beam control is a more mature discipline in optics. Linear phase gradients have been used with holographic optical tweezers to induce particle movement in elongated traps.[110]

6.2.2 Phase Gradient Surfers

Particles floating on the water surface experience a force when exposed to a water-borne sound wave. When this is combined with oblique incidence of the acoustic wave the field will couple to the particle which in turn causes lateral forces that move the object along with the wave. Typically, this effect just leads to particles floating away from the high sound intensity areas. However, the radiation pressure of a focused sound wave causes a deformation of the exposed water surface leading to static crests.[113] The change of curvature across these crests enables

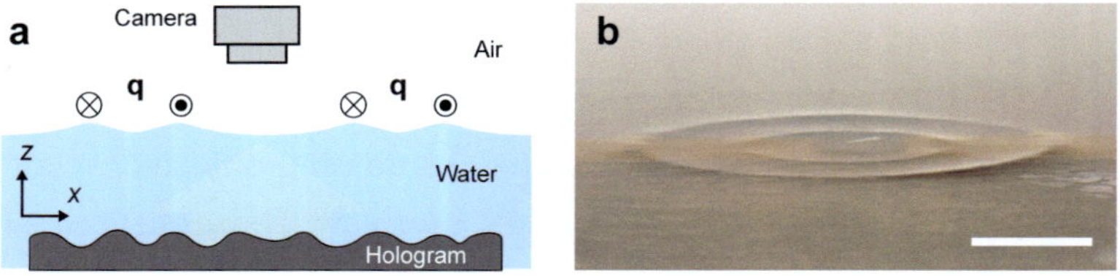

Figure 6.6: (**a**) Schematic of the setup showing crests formed at the water-air interface by two projected concentric rings. The direction of the phase gradient **q** is shown above the crests. (**b**) Oblique angle photograph of the water crests in the experiment. Scale bar, 10 mm. Figure adapted from reference [1].

trapping of objects in a local minimum of capillary energy. Consider Figure 6.6(a), which schematically shows a cut through the experimental setup. There the hologram projects two concentric rings of high sound pressure amplitude that lead to permanent protrusions of the water surface. Floating objects that are positioned on top of a crest are trapped and thereby constrained to only move along the circular track. The projected pressure amplitude thus defines the path that the particle can take. If the hologram further encodes a phase gradient on top of the amplitude, as indicated by the markers, it provides a driving force to push the particle along its path. The photograph in Figure 6.6(b) clearly shows the crests on the water surface for the two concentric rings that are projected by the hologram.

This concept is experimentally demonstrated for two cases—closed and open trajectories. For each case a hologram encodes an image of a curvilinear track of constant pressure amplitude with a superimposed uniform phase gradient of $\sim 1\,\mathrm{rad\,mm^{-1}}$.

Figure 6.7(a,b) show the amplitude and phase maps, respectively, for two concentric circular paths with opposite angular phase gradients.

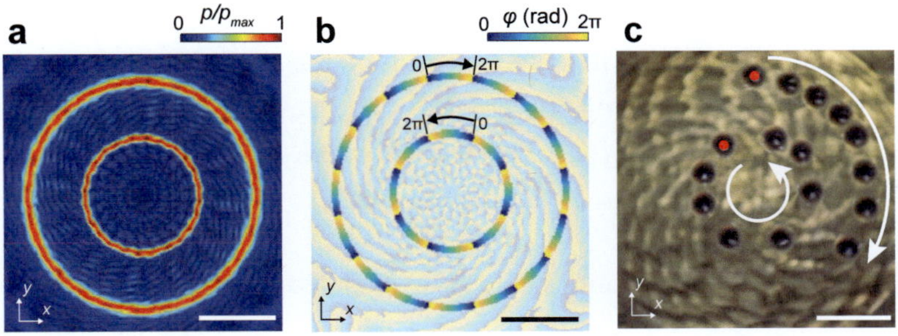

Figure 6.7: Phase gradients on closed tracks. Calculated amplitude (**a**) and phase (**b**) maps of two closed concentric tracks having phase gradients of opposite signs. The areas of high pressure amplitude have been highlighted in the phase map. (**c**) Time-lapse sequence of two objects placed on the rings and moving in opposite directions. All scale bars, 10 mm. Figure adapted from reference [1].

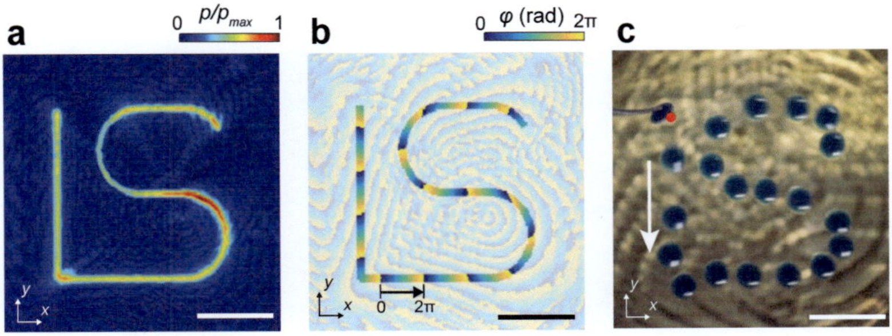

Figure 6.8: Phase gradient on an open track. Calculated amplitude (**a**) and phase (**b**) image of an open track. The areas of high pressure amplitude have been highlighted in the phase map. (**c**) Experimental time-lapse squence of a single object placed at the start of the path (red dot) following the phase gradient along the track. All scale bars, 10 mm. Figure adapted from reference [1].

Particles placed on these rings will follow the paths indefinitely or until the ultrasound transducer is turned off, as seen in the superposition of successive photographs in Figure 6.7(c). These periodic orbits are dependent on the cyclic nature of phase and would be impossible in any system driven by static gradients in acoustic pressure.

The second case is presented in Figure 6.8, where the amplitude and phase maps are shown in panels (a) and (b), respectively. An object that is manually placed at any point of the open track, e.g. the starting point marked by the red dot in panel (c), will follow the entire path until the end. There it slides off the track and floats away.

These experiments demonstrate how acoustic holograms can define a potential map for objects to move without the need for dynamic reconfiguration of the transducer.

Chapter 7

Conclusion

This thesis developed the acoustic hologram as a new approach to shape ultrasound fields with high fidelity. The work on this topic can be divided into (1) computation and implementation of the holographic element, (2) reconstruction and mapping of the sound fields and (3) manipulation of particles and objects via highly structured acoustic fields. Original results have been obtained for all three areas of the thesis. It should be noted, that the methods developed during this thesis are not limited to static printed holograms but will also be useful for phased array transducers (PATs) once those provide sufficient element counts.

The acoustic hologram has been developed in analogy to its optical counterpart, the kinoform,[3] which is a phase element imposing an encoded wavefront onto a passing sound wave. The relative phase lag across the element is related to the local thickness and the difference in speed of sound between hologram and surrounding medium. Given a wave's amplitude distribution and wavefront its whole evolution in space can be predicted via the wave equation. This is the basis of holography and typically the inverse problem is faced, which is the computation of the hologram for a given target field. To this end the iterative angular spectrum approach (IASA) has been adapted to the acoustic domain. This is an algorithm originally proposed for optimization of diffractive optical elements specifically in the near-field. It has proven effective due to its use of the angular spectrum method for wave propagation between hologram and image planes. Only a few tens of iterations are needed to converge to a viable hologram.

To create a physical acoustic hologram the optimized result has to be translated into a thickness profile. Current 3D printers are well-suited to fabricate the intricate and often irregular surface structures with feature sizes down to the 100 μm range, corresponding to the wavelengths of ultrasound in water up to 7.5 MHz. Thus, holograms can be printed for the majority of medical and industrial applications of ultrasound.

Reconstruction of the fields requires only one ultrasound transducer of the right operating frequency and aperture. In this way, the 3D-printed hologram presents a new way to shape acoustic fields with complexities several orders of magnitude higher than conventional methods at a fraction of the cost compared to PAT. The feasibility of the acoustic hologram developed in this thesis has been demonstrated for various different acoustic fields, including single and multiple image planes as well as vortex beams. The reconstructed fields were mapped and validated by calibrated hydrophones. These measurements are very precise but also time-consuming, considering that holograms are capable of encoding whole 3D fields.

This demanded a new method to quickly scan volumetric sound fields. A fast scanning method was developed and presented based on an indirect measurement, which records the temperature evolution of a sound absorbing membrane. Using an infrared radiation (IR) camera a field of 30 mm × 30 mm × 60 mm can be scanned in about 33 s compared to several days using the serial scanning of a hydrophone.

Holographic sound fields were then applied to demonstrate the first assembly of microparticles in truly arbitrary shapes. The particles were arranged in pre-defined 2D patterns on a surface that restricted further motion in the z-direction due to scattering and acoustic streaming. Functionalization of the silicone particles further permitted fixation of the particle assembly triggered by UV illumination. In this process only the particles are bound to each other and the bulk solution stays intact. The trigger signal is not limited to UV light. Acoustic fabrication—the combination of acoustic assembly into objects and subsequent fusion of the structure is another original result of this thesis.

Concluding this work, it has been shown how acoustic holograms present an elegant way to create acoustic fields of arbitrary shape. Those fields can move and arrange microparticles in pre-defined shapes, a first step towards acoustic fabrication. Future work will explore extending this concept to full 3D assembly, possibly requiring multiple transducers in orthogonal or opposite configurations. This step is not trivial and requires new approaches for hologram computation and inclusion of fluid dynamic effects (i.e. acoustic streaming).

A second open research problem is the development of spatial sound modulators in analogy to spatial light modulators. The separation of generation and modulation of a sound wave has clear benefits in terms of complexity and modularity of individual components, e.g. when designing high-power ultrasound transducers. The acoustic hologram has permitted the generation of the most sophisticated sound fields to date and is thus expected to lead to several new applications in the field.

List of References

[1] K. MELDE, A. G. MARK, T. QIU, and P. FISCHER. "Holograms for acoustics". *Nature* **537** (7621): 518–22 (2016). DOI: 10.1038/nature19755

[2] K. MELDE, T. QIU, A. MARK, and P. FISCHER. "Akustische Hologramme steuern Partikel". *Physik in unserer Zeit* **48** (1): 9–10 (2017). DOI: 10.1002/piuz.201790014

[3] L. B. LESEM, P. M. HIRSCH, and J. A. JORDAN. "The Kinoform: A New Wavefront Reconstruction Device". *IBM Journal of Research and Development* **13** (2): 150–5 (1969). DOI: 10.1147/rd.132.0150

[4] B. P. HILDEBRAND and B. B. BRENDEN. *An Introduction to Acoustical Holography.* Plenum Press, New York, 1972. DOI: 10.1007/978-1-4613-4481-0

[5] J. W. GOODMAN. *Introduction to Fourier Optics.* 3rd ed. Roberts & Co. Publishers, Englewood, CO, 2005.

[6] R. PIESTUN and J. SHAMIR. "Synthesis of three-dimensional light fields and applications". *Proceedings of the IEEE* **90** (2): 222–44 (2002). DOI: 10.1109/5.989871

[7] D. GABOR. "A new microscopic principle". *Nature* **161** (4098): 777–8 (1948). DOI: 10.1038/161777a0

[8] U. SCHNARS and W. JUEPTNER. *Digital holography: Digital Hologram Recording, Numerical Reconstruction, and Related Techniques.* Springer-Verlag Berlin Heidelberg, New York, 2005. DOI: 10.1007/b138284

[9] S. MELLIN and G. NORDIN. "Limits of scalar diffraction theory and an iterative angular spectrum algorithm for finite aperture diffractive optical element design". *Optics Express* **8** (13): 705–22 (2001). DOI: 10.1364/OE.8.000705

[10] T. L. SZABO. *Diagnostic Ultrasound Imaging: Inside Out.* 2nd ed. Academic Press, Boston, MA, 2014. DOI: 10.1016/C2011-0-07261-7

[11] B. W. DRINKWATER and P. D. WILCOX. "Ultrasonic arrays for non-destructive evaluation: A review". *NDT & E International* **39** (7): 525–41 (2006). DOI: 10.1016/j.ndteint.2006.03.006

[12] K. MELDE, T. QIU, and P. FISCHER. "Fast spatial scanning of 3D ultrasound fields via thermography". *Applied Physics Letters* **113** (13): 133503 (2018). DOI: 10.1063/1.5046834

[13] LORD RAYLEIGH. "XXXIV. On the pressure of vibrations". *The London, Edinburgh, and Dublin Philosophical Magazine and Journal of Science* **3** (15): 338–46 (2010). DOI: 10.1080/14786440209462769

[14] B. W. DRINKWATER. "Dynamic-field devices for the ultrasonic manipulation of microparticles". *Lab on a Chip* **16** (13): 2360–75 (2016). DOI: 10.1039/C6LC00502K

[15] H. BRUUS. "Acoustofluidics 7: The acoustic radiation force on small particles". *Lab on a Chip* **12** (6): 1014–21 (2012). DOI: 10.1039/c2lc21068a

[16] K. MELDE, E. CHOI, Z. WU, S. PALAGI, T. QIU, and P. FISCHER. "Acoustic Fabrication via the Assembly and Fusion of Particles". *Advanced Materials* **30** (3): 1704507 (2018). DOI: 10.1002/adma.201704507

[17] A. J. HUDSPETH. "How Hearing Happens". *Neuron* **19** (5): 947–50 (1997). DOI: 10.1016/S0896-6273(00)80385-2

[18] P. T. MADSEN and A. SURLYKKE. "Functional convergence in bat and toothed whale biosonars". *Physiology* **28** (5): 276–83 (2013). DOI: 10.1152/physiol.00008.2013

[19] G. JONES and E. C. TEELING. "The evolution of echolocation in bats". *Trends in Ecology and Evolution* **21** (3): 149–56 (2006). DOI: 10.1016/j.tree.2006.01.001

[20] J. J. RASSWEILER, G. G. TAILLY, and C. CHAUSSY. "Progress in Lithotriptor Technology". *European Urology Supplements* **3** (1): 17–36 (2005). DOI: 10.1016/j.euus.2004.11.003

[21] G. TER HAAR. In: "HIFU Tissue Ablation: Concept and Devices". J.-M. ESCOFFRE and A. BOUAKAZ, eds. 3–20. Springer International Publishing, Cham, 2016. DOI: 10.1007/978-3-319-22536-4_1

[22] R. M. S. SIGRIST, J. LIAU, A. E. KAFFAS, M. C. CHAMMAS, and J. K. WILLMANN. "Ultrasound Elastography: Review of Techniques and Clinical Applications". *Theranostics* **7** (5): 1303–29 (2017). DOI: 10.7150/thno.18650

[23] S. MANOHAR and D. RAZANSKY. "Photoacoustics: a historical review". *Advances in Optics and Photonics* **8** (4): 586–617 (2016). DOI: 10.1364/AOP.8.000586

[24] K. S. VALLURU, K. E. WILSON, and J. K. WILLMANN. "Photoacoustic Imaging in Oncology: Translational Preclinical and Early Clinical Experience". *Radiology* **280** (2): 332–49 (2016). DOI: 10.1148/radiol.16151414

[25] F. PETERSSON, L. ABERG, A. M. SWARD-NILSSON, and T. LAURELL. "Free flow acoustophoresis: microfluidic-based mode of particle and cell separation". *Analytical Chemistry* **79** (14): 5117–23 (2007). DOI: 10.1021/ac070444e

[26] E. H. Brandt. “Acoustic physics. Suspended by sound”. *Nature* **413** (6855): 474–5 (2001). DOI: 10.1038/35097192

[27] A. Ozcelik, J. Rufo, F. Guo, Y. Gu, P. Li, J. Lata, and T. J. Huang. “Acoustic tweezers for the life sciences”. *Nature Methods* **15** (12): 1021–28 (2018). DOI: 10.1038/s41592-018-0222-9

[28] S. Palagi, A. G. Mark, S. Y. Reigh, K. Melde, T. Qiu, H. Zeng, C. Parmeggiani, D. Martella, A. Sanchez-Castillo, N. Kapernaum, F. Giesselmann, D. S. Wiersma, E. Lauga, and P. Fischer. “Structured light enables biomimetic swimming and versatile locomotion of photoresponsive soft microrobots”. *Nature Materials* **15** (6): 647–53 (2016). DOI: 10.1038/nmat4569

[29] T. Qiu, S. Palagi, A. G. Mark, K. Melde, F. Adams, and P. Fischer. “Wireless actuation with functional acoustic surfaces”. *Applied Physics Letters* **109** (19): 191602 (2016). DOI: 10.1063/1.4967194

[30] T. Qiu, S. Palagi, A. G. Mark, K. Melde, F. Adams, and P. Fischer. “Active Acoustic Surfaces Enable the Propulsion of a Wireless Robot”. *Advanced Materials Interfaces* **4** (21): 1700933 (2017). DOI: 10.1002/admi.201700933

[31] T. Qiu, F. Adams, S. Palagi, K. Melde, A. Mark, U. Wetterauer, A. Miernik, and P. Fischer. “Wireless Acoustic-Surface Actuators for Miniaturized Endoscopes”. *ACS Applied Materials & Interfaces* **9** (49): 42536–43 (2017). DOI: 10.1021/acsami.7b12755

[32] F. Adams, T. Qiu, A. Mark, K. Melde, S. Palagi, A. Miernik, and P. Fischer. “Wireless micro-robots for endoscopic applications in urology”. *European Urology Supplements* **16** (3): e1914 (2017). DOI: 10.1016/s1569-9056(17)31148-x

[33] R. LERCH, G. SESSLER, and D. WOLF. *Technische Akustik: Grundlagen und Anwendungen.* Springer-Verlag Berlin Heidelberg, 2009. DOI: 10.1007/978-3-540-49833-9

[34] P. M. MORSE and K. U. INGARD. *Theoretical Acoustics.* McGraw-Hill, New York, 1968.

[35] X. CHEN and R. E. APFEL. "Radiation force on a spherical object in an axisymmetric wave field and its application to the calibration of high-frequency transducers". *The Journal of the Acoustical Society of America* **99** (2): 713–24 (1996). DOI: 10.1121/1.414648

[36] O. A. SAPOZHNIKOV, S. A. TSYSAR, V. A. KHOKHLOVA, and W. KREIDER. "Acoustic holography as a metrological tool for characterizing medical ultrasound sources and fields". *The Journal of the Acoustical Society of America* **138** (3): 1515–32 (2015). DOI: 10.1121/1.4928396

[37] D. L. LIU and R. C. WAAG. "Propagation and backpropagation for ultrasonic wavefront design". *IEEE Transactions on Ultrasonics, Ferroelectrics, and Frequency Control* **44** (1): 1–13 (1997). DOI: 10.1109/58.585184

[38] J. J. STAMNES. *Waves in focal regions : Propagation, Diffraction, and Focusing of Light, Sound, and Water Waves.* 1st ed. IOP Publishing Limited, Bristol, UK, 1986.

[39] J. HOLTERMAN and P. GROEN. *An Introduction to Piezoelectric Materials and Components.* Stichting Applied Piezo, Apeldorn, NL, 2012.

[40] J. DUAL and D. MÖLLER. "Acoustofluidics 4: Piezoelectricity and application in the excitation of acoustic fields for ultrasonic particle manipulation". *Lab on a Chip* **12** (3): 506–14 (2012). DOI: 10.1039/c1lc20913b

[41] J. KRAUTKRÄMER and H. KRAUTKRÄMER. *Ultrasonic Testing of Materials.* 4th ed. Springer-Verlag Berlin Heidelberg, New York, 1990. DOI: 10.1007/978-3-662-10680-8

[42] C. H. SHERMAN and J. L. BUTLER. *Transducers and Arrays for Underwater Sound.* 1st ed. Springer, New York, 2007. DOI: 10.1007/978-0-387-33139-3

[43] B. T. KHURI-YAKUB and O. ORALKAN. "Capacitive micromachined ultrasonic transducers for medical imaging and therapy". *Journal of Micromechanics and Microengineering* **21** (5): 54004–14 (2011). DOI: 10.1088/0960-1317/21/5/054004

[44] E. N. LEITH and J. UPATNIEKS. "Reconstructed Wavefronts and Communication Theory". *Journal of the Optical Society of America* **52** (10): 1123–30 (1962). DOI: 10.1364/Josa.52.001123

[45] G. MA, X. FAN, F. MA, J. DE ROSNY, P. SHENG, and M. FINK. "Towards anti-causal Green's function for three-dimensional sub-diffraction focusing". *Nature Physics* **14** (6): 608–12 (2018). DOI: 10.1038/s41567-018-0082-3

[46] J. R. FIENUP. "Phase retrieval algorithms: a comparison". *Applied Optics* **21** (15): 2758–69 (1982). DOI: 10.1364/AO.21.002758

[47] R. W. GERCHBERG and W. O. SAXTON. "A Practical Algorithm for the Determination of Phase from Image and Diffraction Plane Pictures". *Optik* **35** (2): 237–46 (1972).

[48] R. DI LEONARDO, F. IANNI, and G. RUOCCO. "Computer generation of optimal holograms for optical trap arrays". *Optics Express* **15** (4): 1913–22 (2007). DOI: 10.1364/OE.15.001913

[49] H. H. BAUSCHKE, P. L. COMBETTES, and D. R. LUKE. “Phase retrieval, error reduction algorithm, and Fienup variants: a view from convex optimization”. *Journal of the Optical Society of America A* **19** (7): 1334–45 (2002). DOI: 10.1364/JOSAA.19.001334

[50] M. MEISTER and R. J. WINFIELD. “Novel approaches to direct search algorithms for the design of diffractive optical elements”. *Optics Communications* **203** (1-2): 39–49 (2002). DOI: 10.1016/S0030-4018(01)01741-2

[51] K. MATSUSHIMA and T. SHIMOBABA. “Band-limited angular spectrum method for numerical simulation of free-space propagation in far and near fields”. *Optics Express* **17** (22): 19662–73 (2009). DOI: 10.1364/OE.17.019662

[52] L. XU, X. PENG, Z. GUO, J. MIAO, and A. ASUNDI. “Imaging analysis of digital holography”. *Optics Express* **13** (7): 2444–52 (2005). DOI: 10.1364/OPEX.13.002444

[53] R. PIESTUN and D. A. B. MILLER. “Electromagnetic degrees of freedom of an optical system”. *The Journal of the Optical Society of America A* **17** (5): 892–902 (2000). DOI: 10.1364/JOSAA.17.000892

[54] Y. HERTZBERG and G. NAVON. “Bypassing absorbing objects in focused ultrasound using computer generated holographic technique”. *Medical Physics* **38** (12): 6407–15 (2011). DOI: 10.1118/1.3651464

[55] A. NEILD. “Acoustics: Motion controlled by sound”. *Nature* **537** (7621): 493 (2016). DOI: 10.1038/537493a

[56] D. A. BURALLI, G. M. MORRIS, and J. R. ROGERS. “Optical performance of holographic kinoforms”. *Applied Optics* **28** (5): 976–83 (1989). DOI: 10.1364/AO.28.000976

[57] M. D. Brown, B. T. Cox, and B. E. Treeby. "Design of multi-frequency acoustic kinoforms". *Applied Physics Letters* **111** (24): 244101–5 (2017). DOI: 10.1063/1.5004040

[58] S. Gspan, A. Meyer, S. Bernet, and M. Ritsch-Marte. "Optoacoustic generation of a helicoidal ultrasonic beam". *The Journal of the Acoustical Society of America* **115** (3): 1142–6 (2004). DOI: 10.1121/1.1643367

[59] M. D. Brown, D. I. Nikitichev, B. E. Treeby, and B. T. Cox. "Generating arbitrary ultrasound fields with tailored optoacoustic surface profiles". *Applied Physics Letters* **110** (9): 094102 (2017). DOI: 10.1063/1.4976942

[60] A. Meyer, S. Gspan, S. Bernet, and M. Ritsch-Marte. "Binary optoacoustic holography with a spatial light modulator". *Journal of Applied Physics* **96** (10): 5886–91 (2004). DOI: 10.1063/1.1801162

[61] *VeroClear - Rigid Transparent Polyjet Material.* Datasheet. Stratasys Ltd., USA, 2018

[62] G. Sinclair, J. Leach, P. Jordan, G. Gibson, E. Yao, Z. Laczik, M. Padgett, and J. Courtial. "Interactive application in holographic optical tweezers of a multi-plane Gerchberg-Saxton algorithm for three-dimensional light shaping". *Optics Express* **12** (8): 1665–70 (2004). DOI: 10.1364/OPEX.12.001665

[63] C. E. Demore, Z. Yang, A. Volovick, S. Cochran, M. P. MacDonald, and G. C. Spalding. "Mechanical evidence of the orbital angular momentum to energy ratio of vortex beams". *Physical Review Letters* **108** (19): 194301 (2012). DOI: 10.1103/PhysRevLett.108.194301

[64] E. Martin, E. Z. Zhang, J. A. Guggenheim, P. C. Beard, and B. E. Treeby. "Rapid Spatial Mapping of Focused Ultrasound Fields Using a Planar Fabry-Perot Sensor". *IEEE Transactions on Ultrasonics, Ferroelectrics, and Frequency Control* **64** (11): 1711–22 (2017). DOI: 10.1109/TUFFC.2017.2748886

[65] J. Seo, N. Koizumi, K. Yoshinaka, N. Sugita, A. Nomiya, Y. Homma, Y. Matsumoto, and M. Mitsuishi. "Three-dimensional computer-controlled acoustic pressure scanning and quantification of focused ultrasound". *IEEE Transactions on Ultrasonics, Ferroelectrics, and Frequency Control* **57** (4): 883–91 (2010). DOI: 10.1109/TUFFC.2010.1492

[66] J. D. Maynard, E. G. Williams, and Y. Lee. "Nearfield acoustic holography: I. Theory of generalized holography and the development of NAH". *The Journal of the Acoustical Society of America* **78** (4): 1395–1413 (1985). DOI: 10.1121/1.392911

[67] A. Korpel. "Acousto-optics—A review of fundamentals". *Proceedings of the IEEE* **69** (1): 48–53 (1981). DOI: 10.1109/PROC.1981.11919

[68] E. Zhang, J. Laufer, and P. Beard. "Backward-mode multiwavelength photoacoustic scanner using a planar Fabry-Perot polymer film ultrasound sensor for high-resolution three-dimensional imaging of biological tissues". *Applied Optics* **47** (4): 561–77 (2008). DOI: 10.1364/Ao.47.000561

[69] Y. R. Jia, Q. Wei, D. J. Wu, Z. Xu, and X. J. Liu. "Generation of fractional acoustic vortex with a discrete Archimedean spiral structure plate". *Applied Physics Letters* **112** (17): 173501 (2018). DOI: 10.1063/1.5026646

[70] A. R. HARLAND, J. N. PETZING, and J. R. TYRER. "Nonperturbing measurements of spatially distributed underwater acoustic fields using a scanning laser Doppler vibrometer". *The Journal of the Acoustical Society of America* **115** (1): 187–95 (2004). DOI: 10.1121/1.1635841

[71] T. NAKAMURA, R. IWASAKI, S. YOSHIZAWA, and S. UMEMURA. "Quantitative measurement of ultrasonic pressure field using combination of optical phase contrast and nonlinear acoustic holography methods". *Japanese Journal of Applied Physics* **57** (7S1): 07lb13 (2018). DOI: 10.7567/jjap.57.07lb13

[72] K. MARTIN and R. FERNANDEZ. "A thermal beam-shape phantom for ultrasound physiotherapy transducers". *Ultrasound in Medicine & Biology* **23** (8): 1267–74 (1997). DOI: 10.1016/S0301-5629(97)00109-9

[73] G. A. LOPEZ MUNOZ and G. A. VALENTINO OROZCO. In: "Three Dimensional Temperature Distribution Analysis of Ultrasound Therapy Equipments Using Thermochromic Liquid Crystal Films". G. TKACHENKO, ed. Chap. 5. IntechOpen, 2009. DOI: 10.5772/9689

[74] F. P. CURRA, P. D. MOURAD, V. A. KHOKHLOVA, R. O. CLEVELAND, and L. A. CRUM. "Numerical simulations of heating patterns and tissue temperature response due to high-intensity focused ultrasound". *IEEE Transactions on Ultrasonics, Ferroelectrics, and Frequency Control* **47** (4): 1077–89 (2000). DOI: 10.1109/58.852092

[75] M. NECATI ÖZIŞIK. *Heat Conduction*. 2nd. John Wiley & Sons, Inc., New York, 1993.

[76] K. Olofsson, B. Hammarström, and M. Wiklund. "Ultrasonic Based Tissue Modelling and Engineering". *Micromachines* **9** (11): 594 (2018). DOI: 10.3390/mi9110594

[77] Z. Ma, D. J. Collins, J. Guo, and Y. Ai. "Mechanical Properties Based Particle Separation via Traveling Surface Acoustic Wave". *Analytical Chemistry* **88** (23): 11844–51 (2016). DOI: 10.1021/acs.analchem.6b03580

[78] C. R. P. Courtney, C. K. Ong, B. W. Drinkwater, A. L. Bernassau, P. D. Wilcox, and D. R. S. Cumming. "Manipulation of particles in two dimensions using phase controllable ultrasonic standing waves". *Proceedings of the Royal Society A* **468** (2138): 337–60 (2011). DOI: 10.1098/rspa.2011.0269

[79] I. Leibacher, W. Dietze, P. Hahn, J. Wang, S. Schmitt, and J. Dual. "Acoustophoresis of hollow and core-shell particles in two-dimensional resonance modes". *Microfluidics and Nanofluidics* **16** (3): 513–24 (2014). DOI: 10.1007/s10404-013-1240-7

[80] M. Caleap and B. W. Drinkwater. "Acoustically trapped colloidal crystals that are reconfigurable in real time". *Proceedings of the National Academy of Sciences of the United States of America* **111** (17): 6226–30 (2014). DOI: 10.1073/pnas.1323048111

[81] C. R. P. Courtney, C. E. M. Demore, H. X. Wu, A. Grinenko, P. D. Wilcox, S. Cochran, and B. W. Drinkwater. "Independent trapping and manipulation of microparticles using dexterous acoustic tweezers". *Applied Physics Letters* **104** (15): 154103 (2014). DOI: 10.1063/1.4870489

[82] X. CHEN, K. H. LAM, R. CHEN, Z. CHEN, P. YU, Z. CHEN, K. K. SHUNG, and Q. ZHOU. "An adjustable multi-scale single beam acoustic tweezers based on ultrahigh frequency ultrasonic transducer". *Biotechnology and Bioengineering* **114** (11): 2637–47 (2017). DOI: 10.1002/bit.26365

[83] J. LEE, S. Y. TEH, A. LEE, H. H. KIM, C. LEE, and K. K. SHUNG. "Single beam acoustic trapping". *Applied Physics Letters* **95** (7): 73701 (2009). DOI: 10.1063/1.3206910

[84] D. BARESCH, J. L. THOMAS, and R. MARCHIANO. "Observation of a Single-Beam Gradient Force Acoustical Trap for Elastic Particles: Acoustical Tweezers". *Physical Review Letters* **116** (2): 024301 (2016). DOI: 10.1103/PhysRevLett.116.024301

[85] A. FRANKLIN, A. MARZO, R. MALKIN, and B. W. DRINKWATER. "Three-dimensional ultrasonic trapping of micro-particles in water with a simple and compact two-element transducer". *Applied Physics Letters* **111** (9): 094101 (2017). DOI: 10.1063/1.4992092

[86] M. S. SCHOLZ, B. W. DRINKWATER, T. M. LLEWELLYN-JONES, and R. S. TRASK. "Counterpropagating wave acoustic particle manipulation device for the effective manufacture of composite materials". *IEEE Transactions on Ultrasonics, Ferroelectrics, and Frequency Control* **62** (10): 1845–55 (2015). DOI: 10.1109/TUFFC.2015.007116

[87] J. GREENHALL and B. RAEYMAEKERS. "3D Printing Macroscale Engineered Materials Using Ultrasound Directed Self-Assembly and Stereolithography". *Advanced Materials Technologies* **2** (9): 1700122 (2017). DOI: 10.1002/Admt.201700122

[88] R. T. BEYER. “Radiation pressure—the history of a mislabeled tensor”. *The Journal of the Acoustical Society of America* **63** (4): 1025–30 (1978). DOI: 10.1121/1.381833

[89] V. F. HUMPHREY. “Ultrasound and matter–physical interactions”. *Progress in Biophysics & Molecular Biology* **93** (1-3): 195–211 (2007). DOI: 10.1016/j.pbiomolbio.2006.07.024

[90] T. HASEGAWA. “Comparison of two solutions for acoustic radiation pressure on a sphere”. *The Journal of the Acoustical Society of America* **61** (6): 1445–8 (1977). DOI: 10.1121/1.381460

[91] L. P. GOR’KOV. “On the forces acting on a small particle in an acoustical field in an ideal fluid”. *Soviet Physics Doklady* **6** (9): 773–5 (1962).

[92] O. A. SAPOZHNIKOV and M. R. BAILEY. “Radiation force of an arbitrary acoustic beam on an elastic sphere in a fluid”. *The Journal of the Acoustical Society of America* **133** (2): 661–76 (2013). DOI: 10.1121/1.4773924

[93] R. H. PRITCHARD, P. LAVA, D. DEBRUYNE, and E. M. TERENTJEV. “Precise determination of the Poisson ratio in soft materials with 2D digital image correlation”. *Soft Matter* **9** (26): 6037–45 (2013). DOI: 10.1039/c3sm50901j

[94] M. SINHA and D. J. BUCKLEY. In: “Acoustic Properties of Polymers”. J. E. MARK, ed. Chap. 60, 1021–31. Springer, New York, 2007. DOI: 10.1007/978-0-387-69002-5

[95] L. M. JOHNSON, L. GAO, IV CW SHIELDS, M. SMITH, K. EFIMENKO, K. CUSHING, J. GENZER, and G. P. LOPEZ. “Elastomeric microparticles for acoustic mediated bioseparations”. *Journal of Nanobiotechnology* **11** (22): (2013). DOI: 10.1186/1477-3155-11-22

[96] J. ZHOU, A. V. ELLIS, and N. H. VOELCKER. "Recent developments in PDMS surface modification for microfluidic devices". *Electrophoresis* **31** (1): 2–16 (2010). DOI: 10.1002/elps.200900475

[97] S. HU, X. REN, M. BACHMAN, C. E. SIMS, G. P. LI, and N. L. ALLBRITTON. "Surface-directed, graft polymerization within microfluidic channels". *Analytical Chemistry* **76** (7): 1865–70 (2004). DOI: 10.1021/ac049937z

[98] V. MUCCI and C. VALLO. "Efficiency of 2,2-dimethoxy-2-phenylacetophenone for the photopolymerization of methacrylate monomers in thick sections". *Journal of Applied Polymer Science* **123** (1): 418–25 (2012). DOI: 10.1002/app.34473

[99] J. N. LEE, C. PARK, and G. M. WHITESIDES. "Solvent compatibility of poly-dimethylsiloxane-based microfluidic devices". *Analytical Chemistry* **75** (23): 6544–54 (2003). DOI: 10.1021/ac0346712

[100] K. BÜCKS and H. MÜLLER. "Über einige Beobachtungen an schwingenden Piezoquarzen und ihrem Schallfeld". *Zeitschrift für Physik* **84** (1-2): 75–86 (1933). DOI: 10.1007/BF01330275

[101] V. VANDAELE, P. LAMBERT, and A. DELCHAMBRE. "Non-contact handling in microassembly: Acoustical levitation". *Precision Engineering* **29** (4): 491–505 (2005). DOI: 10.1016/j.precisioneng.2005.03.003

[102] S. SANTESSON and S. NILSSON. "Airborne chemistry: acoustic levitation in chemical analysis". *Analytical and Bioanalytical Chemistry* **378** (7): 1704–9 (2004). DOI: 10.1007/s00216-003-2403-2

[103] E. H. TRINH. "Compact acoustic levitation device for studies in fluid dynamics and material science in the laboratory and microgravity". *Review of Scientific Instruments* **56** (11): 2059–65 (1985). DOI: 10.1063/1.1138419

[104] D. FORESTI, M. NABAVI, M. KLINGAUF, A. FERRARI, and D. POULIKAKOS. "Acoustophoretic contactless transport and handling of matter in air". *Proceedings of the National Academy of Sciences of the United States of America* **110** (31): 12549–54 (2013). DOI: 10.1073/pnas.1301860110

[105] W. J. XIE and B. WEI. "Parametric study of single-axis acoustic levitation". *Applied Physics Letters* **79** (6): 881–3 (2001). DOI: 10.1063/1.1391398

[106] W. J. XIE, C. D. CAO, Y. J. LÜ, Z. Y. HONG, and B. WEI. "Acoustic method for levitation of small living animals". *Applied Physics Letters* **89** (21): 214102 (2006). DOI: 10.1063/1.2396893

[107] Y. OCHIAI, T. HOSHI, and J. REKIMOTO. "Three-dimensional mid-air acoustic manipulation by ultrasonic phased arrays". *PLoS ONE* **9** (5): e97590 (2014). DOI: 10.1371/journal.pone.0097590

[108] A. MARZO, S. A. SEAH, B. W. DRINKWATER, D. R. SAHOO, B. LONG, and S. SUBRAMANIAN. "Holographic acoustic elements for manipulation of levitated objects". *Nature Communications* **6**, 8661 (2015). DOI: 10.1038/ncomms9661

[109] A. MARZO and B. W. DRINKWATER. "Holographic acoustic tweezers". *Proceedings of the National Academy of Sciences of the United States of America* **116** (1): 84–9 (2019). DOI: 10.1073/pnas.1813047115

[110] Y. ROICHMAN, B. SUN, Y. ROICHMAN, J. AMATO-GRILL, and D. G. GRIER. "Optical forces arising from phase gradients". *Physical Review Letters* **100** (1): 013602 (2008). DOI: 10.1103/PhysRevLett.100.013602

[111] A. ANHAUSER, R. WUNENBURGER, and E. BRASSELET. "Acoustic rotational manipulation using orbital angular momentum transfer". *Physical Review Letters* **109** (3): 034301 (2012). DOI: 10.1103/PhysRevLett.109.034301

[112] D. BARESCH, J. L. THOMAS, and R. MARCHIANO. "Orbital Angular Momentum Transfer to Stably Trapped Elastic Particles in Acoustical Vortex Beams". *Physical Review Letters* **121** (7): 074301 (2018). DOI: 10.1103/PhysRevLett.121.074301

[113] B. ISSENMANN, R. WUNENBURGER, H. CHRAIBI, M. GANDIL, and J. P. DELVILLE. "Unsteady deformations of a free liquid surface caused by radiation pressure". *Journal of Fluid Mechanics* **682**, 460–90 (2011). DOI: 10.1017/jfm.2011.236

Acknowledgements

First and foremost I want to thank my advisor Prof. Peer Fischer for the opportunity to work on this topic and for creating and managing such a positive and inspiring work environment. Further, I would like to express my gratitude to Prof. Jan Korvink for supporting me as an external doctoral student at the Institute of Microstructure Technology and to Prof. Wolfgang Seemann for chairing the examination committee of this dissertation. Thank you for kindly agreeing to take part in the evaluation of this thesis.

Andrew Mark has been a great scientific mentor for me and I would like to thank him for that and the curious questions involving my work and also for showing me how to structure and write a research article. On the same note, I would like to thank Tian Qiu and Stefano Palagi, who have always been very helpful in discussing ideas and solving technical issues with the setup or the simulations. On top of that, Tian has also taken good care of the 3D printer and he's printed literally hundreds of holograms for me. A special thank you goes to Eunjin Choi for her creative ideas with functionalizing the silicone particles and her assistance during the acoustic fabrication experiments.

Thanks to our collaborators Peter Weber and Daniel Schmitt from the Fraunhofer IBMT for supplying the first piezo plates and for their technical advice regarding transducer fabrication and ultrasound applications.

A big thank you to the past and present members of the Micro, Nano and Molecular Systems lab for the amazing working atmosphere, for their support and friendship: Johannes Sachs, Mariana Alarcon-Correa,

Vincent Kadiri, Alejandro Posada, Jan-Philipp Günther, Tingting Yu, Nico Segreto, Jonas Troll, Björn Miksch, Melanie Adams, Alexander von Rohr, Stefan Knoppe, Angela Holst-Jahn, Ziggy Wu, Korbinian Pöppel, Zhichao Ma, Hannah-Noa Barad, Hyunah Kwon, Hyeon-Ho Jeong.

My sincere thanks to Jutta Hess for the constant administrative support and keeping the group organized.

Finally, I would like to thank my family for all their love and support throughout the years.